Collins

AQA GCSE 9-1 ...on

Combined Science

Trilogy

Combined Science

Higher

AQA GCSE 9-1

Workbook

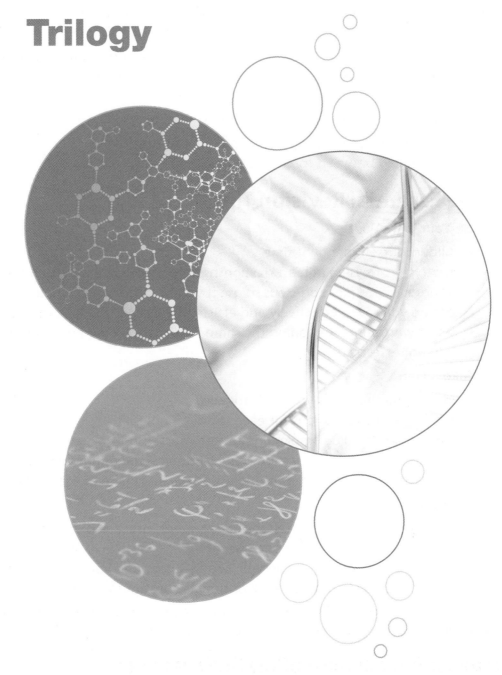

Ian Honeysett, Emma Poole,
Gemma Young and Nathan Goodman

Revision Tips

Rethink Revision

Have you ever taken part in a quiz and thought *'I know this!'* but, despite frantically racking your brain, you just couldn't come up with the answer?

It's very frustrating when this happens but, in a fun situation, it doesn't really matter. However, in your GCSE exams, it will be essential that you can recall the relevant information quickly when you need to.

Most students think that revision is about making sure you **know** *stuff*. Of course, this is important, but it is also about becoming confident that you can **retain** that *stuff* over time and **recall** it quickly when needed.

Revision That Really Works

Experts have discovered that there are two techniques that help with all of these things and consistently produce better results in exams compared to other revision techniques.

Applying these techniques to your GCSE revision will ensure you get better results in your exams and will have all the relevant knowledge at your fingertips when you start studying for further qualifications, like AS and A Levels, or begin work.

It really isn't rocket science either – you simply need to:

- **test yourself** on each topic as many times as possible
- **leave a gap** between the test sessions.

Three Essential Revision Tips

1. **Use Your Time Wisely**

    - Allow yourself plenty of time.
    - Try to start revising at least six months before your exams – it's more effective and less stressful.
    - Your revision time is precious so use it wisely – using the techniques described on this page will ensure you revise effectively and efficiently and get the best results.
    - Don't waste time re-reading the same information over and over again – it's time-consuming and not effective!

2. **Make a Plan**

    - Identify all the topics you need to revise.
    - Plan at least five sessions for each topic.
    - One hour should be ample time to test yourself on the key ideas for a topic.
    - Spread out the practice sessions for each topic – the optimum time to leave between each session is about one month but, if this isn't possible, just make the gaps as big as realistically possible.

3. **Test Yourself**

    - Methods for testing yourself include: quizzes, practice questions, flashcards, past papers, explaining a topic to someone else, etc.
    - Don't worry if you get an answer wrong – provided you check what the correct answer is, you are more likely to get the same or similar questions right in future!

Visit our website for more information about the benefits of these techniques and for further guidance on how to plan ahead and make them work for you.

www.collins.co.uk/collinsGCSErevision

Contents

Biology Paper 1   **Cell Biology**

Cell Structure   6
Investigating Cells   7
Cell Division   8
Transport In and Out of Cells   9

Biology Paper 1   **Organisation**

Levels of Organisation   10
Digestion   11
Blood and the Circulation   12
Non-Communicable Diseases   13
Transport in Plants   14

Biology Paper 1   **Infection and Response**

Pathogens and Disease   15
Human Defences Against Disease   16
Treating Diseases   17

Biology Paper 1   **Bioenergetics**

Photosynthesis   18
Respiration and Exercise   19

Biology Paper 2   **Homeostasis and Response**

Homeostasis and the Nervous System   20
Hormones and Homeostasis   21
Hormones and Reproduction   22

Biology Paper 2   **Inheritance, Variation and Evolution**

Sexual and Asexual Reproduction   23
Patterns of Inheritance   24
Variation and Evolution   25
Manipulating Genes   26
Classification   27

Biology Paper 2   **Ecology**

Ecosystems   28
Cycles and Feeding Relationships   29
Disrupting Ecosystems   30

Chemistry Paper 1 **Atomic Structure and the Periodic Table**

Atoms, Elements, Compounds and Mixtures   31
Atoms and the Periodic Table   32
The Periodic Table   33

Chemistry Paper 1 **Bonding, Structure and Properties of Matter**

States of Matter   34
Ionic Compounds   35
Covalent Compounds   36
Metals and Special Materials   37

Contents

Chemistry Paper 1 Quantitative Chemistry

Conservation of Mass 38
Amount of Substance 39

Chemistry Paper 1 Chemical Changes

Reactivity of Metals 40
The pH Scale and Salts 41
Electrolysis 42

Chemistry Paper 1 Energy Changes

Exothermic and Endothermic Reactions 43
Measuring Energy Changes 44

Chemistry Paper 2 The Rate and Extent of Chemical Change

Rate of Reaction 45
Reversible Reactions 46

Chemistry Paper 2 Organic Chemistry

Alkanes 47
Cracking Hydrocarbons 48

Chemistry Paper 2 Chemical Analysis

Chemical Analysis 49

Chemistry Paper 2 Chemistry of the Atmosphere

The Earth's Atmosphere 50
Greenhouse Gases 51

Chemistry Paper 2 Using Resources

Earth's Resources 52
Using Resources 53

Physics Paper 2 Forces

Forces – An Introduction 54
Forces in Action 55
Forces and Motion 56
Forces and Acceleration 57
Terminal Velocity and Momentum 58
Stopping and Braking 59

Physics Paper 1 Energy

Energy Stores and Transfers 60
Energy Transfers and Resources 61

Physics Paper 2 Waves

Waves and Wave Properties 62
Electromagnetic Waves 63
The Electromagnetic Spectrum 64

Contents

Physics Paper 1 **Electricity**

An Introduction to Electricity   65
Circuits and Resistance   66
Circuits and Power   67
Domestic Uses of Electricity   68
Electrical Energy in Devices   69

Physics Paper 2 **Magnetism and Electromagnetism**

Magnetism and Electromagnetism   70
The Motor Effect   71

Physics Paper 1 **Particle Model of Matter**

Particle Model of Matter   72

Physics Paper 1 **Atomic Structure**

Atoms and Isotopes   73
Nuclear Radiation   74
Half-Life   75

Biology Paper 1 **GCSE Combined Science Practice Exam Paper**   **77**

Biology Paper 2 **GCSE Combined Science Practice Exam Paper**   **93**

Chemistry Paper 1 **GCSE Combined Science Practice Exam Paper**   **109**

Chemistry Paper 2 **GCSE Combined Science Practice Exam Paper**   **125**

Physics Paper 1 **GCSE Combined Science Practice Exam Paper**   **141**

Physics Paper 2 **GCSE Combined Science Practice Exam Paper**   **155**

Physics Equation Sheet   **169**

Periodic Table   **170**

Answers   **171**

Cell Structure

1 **Figure 1** shows an animal cell.

   Which part of an animal cell do the following descriptions refer to?
   Write the letter **A**, **B** or **C** from the diagram next to each statement.

   Figure 1

   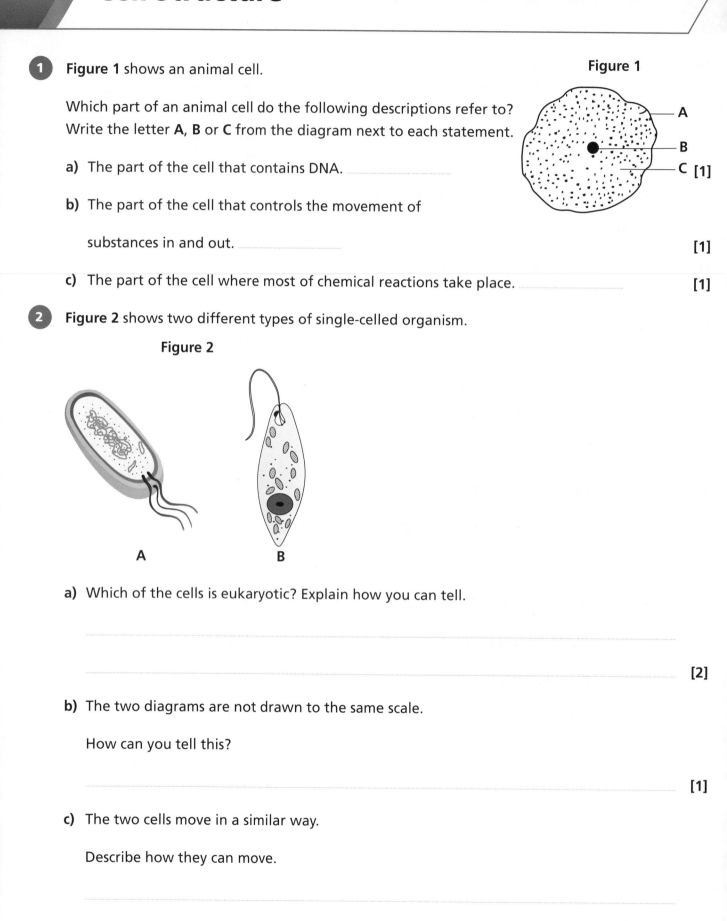

   a) The part of the cell that contains DNA. .... [1]

   b) The part of the cell that controls the movement of

       substances in and out. .... [1]

   c) The part of the cell where most of chemical reactions take place. .... [1]

2 **Figure 2** shows two different types of single-celled organism.

   Figure 2

   A   B

   a) Which of the cells is eukaryotic? Explain how you can tell.

       ....

       .... [2]

   b) The two diagrams are not drawn to the same scale.

       How can you tell this?

       .... [1]

   c) The two cells move in a similar way.

       Describe how they can move.

       ....

       .... [2]

   Total Marks .... / 8

Investigating Cells

1 Rajesh uses his light microscope to view and draw a cheek cell.
He puts some cheek cells on to a slide and adds a few drops of a blue solution.

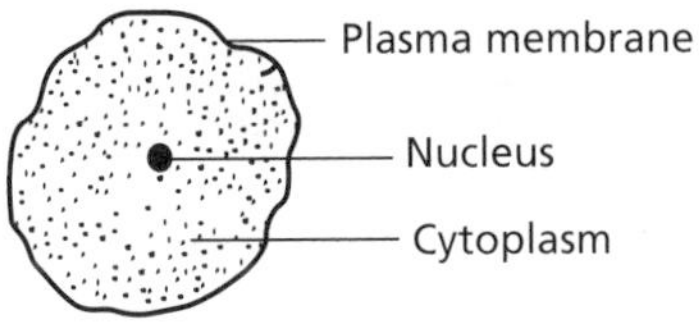

Figure 1 shows what Rajesh draws.

Figure 1

a) Why does Rajesh add a few drops of blue solution to the cells?

....

.... **[2]**

b) A cheek cell is actually 0.03mm in diameter.

   i) What is the diameter in micrometres?

   Answer: .... **[1]**

   ii) Calculate the magnification of Rajesh's drawing.

   $$\text{magnification} = \frac{\text{size of image}}{\text{size of real object}}$$

   Answer: .... **[2]**

2 Roger is using a light microscope to look at some onion cells on a slide.
On the slide there is also a scale.
Each division on the scale represents **0.01mm**.

Figure 2 shows what Roger sees.

Figure 2

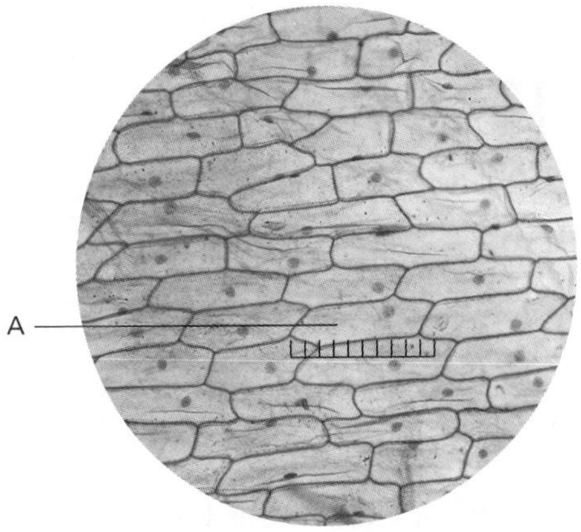

a) Estimate the length of cell **A** in **Figure 2** in millimetres.

   Answer: .... **[1]**

b) How wide is the nucleus of an onion cell?
   Tick **one** box next to the best estimate.

   100 micrometres ☐   1 micrometre ☐

   10 micrometres ☐   0.1 micrometres ☐ **[1]**

   Total Marks .... / 7

Cell Division

1 A student investigates the growth of onions.
He puts an onion bulb in a jar of water.
The bulb starts to grow roots.

a) Cells in the root tip are dividing.

Which part of the cell cycle involves cells dividing?

.... **[1]**

b) The student makes a slide of the onion root and looks at it with a light microscope.
He sees chromosomes inside some of the cells.

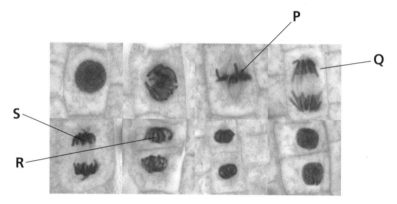

Cells **P, Q, R** and **S** are in different stages of cell division.

Put the stages in order.
One has been done for you.

		S	

[2]

2 Read the following quote and answer the questions below.

"People are always against new ideas such as using stem cells, but within a few years they
will be used all the time to cure diseases."

a) How can stem cells be used to cure diseases?

....

.... **[2]**

b) Give **one** reason why people may be against using stem cells to cure diseases.

.... **[1]**

Total Marks .... / 6

Transport In and Out of Cells

1 The three blocks in **Figure 1** were cut from a block of agar jelly that had been dyed green.

Figure 1

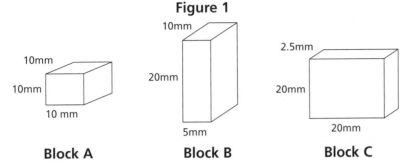

Block A

Block B

Block C

a) Calculate the surface area of block **A**. Put the results in **Table 1**.

Table 1

Block	Surface Area (mm²)	Volume (mm³)
A		1000
B	700	1000
C	1000	1000

[1]

b) The green dye turns yellow when acid moves through into the agar jelly.

What process would cause the acid to move through the agar jelly?

Answer: _____ [1]

c) The three blocks are submerged in an acid solution.

Which block would be the first to change colour completely? Explain your answer.

_____

_____

_____ [3]

2 Give **two** differences between:

a) Osmosis and diffusion.

_____

_____ [2]

b) Diffusion and active transport.

_____

_____ [2]

Total Marks _____ / 9

Levels of Organisation

1 **Systems**, **cells**, **organs** and **tissues** are all levels of organisation in living organisms.

a) Write down these four levels in order of complexity, with the least complex first.

.... **[1]**

b) Give the level of organisation of each of these structures.
The first one has been done for you.

A lymphocyte	*cell*
The heart	
A neurone	
A leaf	
Epithelia on the skin	

[4]

c) **Table 1** shows the number of mitochondria in different types of cell.

Which statement could best explain the data in the table?
Tick **one** box.

Table 1

Type of Cell	Number of Mitochondria
liver	1500
heart muscle	5000
skin	100

Heart muscle cells are specialised to make protein. ☐

Skin cells need large amounts of energy. ☐

Liver cells do not require much energy. ☐

Muscle cells are specialised for contraction. ☐ **[1]**

2 **Figure 1** shows phloem and xylem cells.

Describe how phloem and xylem cells are specialised for their functions in plants.

....

....

....

....

Figure 1

Phloem

Xylem

[4]

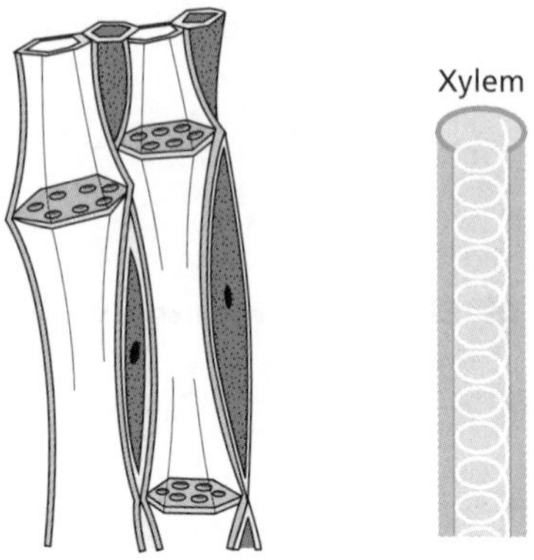

Total Marks .... / 10

Digestion

1 The enzyme amylase breaks down starch.

The graph in **Figure 1** shows the effect of temperature on the enzyme amylase.

Figure 1

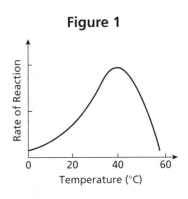

a) Describe the pattern shown in the graph.

....

.... **[2]**

b) Give the optimum temperature of this enzyme.

.... **[1]**

2 Lipids (fats) are digested in the body by the enzyme lipase.

a) Give one place in the body where lipase is made.

Answer: .... **[1]**

b) Some people want to be able to eat foods containing fat without gaining weight.
One way to do this is to replace fats in food with a substance called olestra.
Figure 2 shows an olestra molecule.

Figure 2

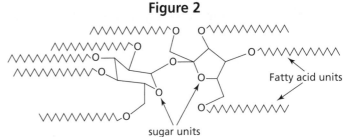

i) Write down **one** difference between an olestra molecule and a normal fat molecule.

.... **[1]**

ii) Suggest why lipase cannot digest olestra.

....

.... **[2]**

Total Marks .... / 7

Blood and the Circulation

1   Match the numbers on **Figure 1** with the blood vessels listed below.
    Write the appropriate numbers in the boxes provided.

Figure 1

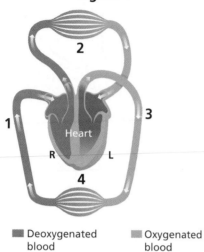

■ Deoxygenated   ■ Oxygenated
   blood   blood

Aorta   ☐

Vena cava   ☐

Capillaries in the lungs   ☐

Capillaries in the body   ☐   **[3]**

2   Haemoglobin is found in red blood cells.

   a) Explain how haemoglobin supplies the tissues of the body with oxygen.

   ....

   ....

   ....   **[3]**

   b) Some people have a condition called sickle-cell anaemia.
        This affects their haemoglobin and can make their red
        blood cells change shape.

        Explain why the red blood cells do **not** work so
        well after they change shape.

Figure 2

Normal red   Sickled red
blood cell   blood cell

   ....

   ....   **[2]**

Total Marks .... / 8

Non-Communicable Diseases

1 A build-up of fatty material in the wall of the coronary arteries of the heart is called atheroma or atherosclerosis.

It has been known for many years that atheromas can cause a heart attack.

a) i) What is a heart attack?

_____ [1]

ii) How can an atheroma cause a heart attack?

_____

_____ [2]

b) People used to think that the only cause of heart attacks was too much saturated fat, causing atheromas.

A scientist looked at groups of men who did different jobs involving different amounts of physical activity.

He recorded how many of these men suffered from heart attacks and how many had large atheromas. His results are shown on the graphs in **Figure 1**.

Figure 1

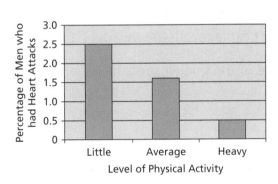

What conclusions can you make from these results?

_____

_____

_____ [3]

Total Marks _____ / 6

Transport in Plants

1 A student sets up an investigation into water transport in a plant as shown in **Figure 1**.
She measures the mass of the test tube and contents at the start and
again the next day.

Figure 1

She repeats the experiment with the plant under different conditions.

The student's results are shown in **Table 1**.

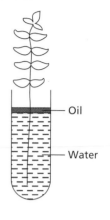

Table 1

Conditions	Mass of Tube and Contents (g)	
	At Start	After 1 Day
normal	29.7	26.5
windy	30.4	25.5
humid	29.2	27.2

a) Complete the following sentences about the student's results.

The plant loses mass because it loses .... by a process called

....

The greatest mass is lost from the plant in .... conditions.   **[3]**

b) Predict what the student's results would be under the following conditions compared
with the normal conditions.
You must explain your answers.

   i) Some of the smaller roots have been removed from the plant.

   ....

   ....   **[2]**

   ii) Some of the leaves are painted with nail varnish.

   ....

   ....   **[2]**

2 Give **two** differences between the movement of water and the movement of sugars in a plant.

....

....   **[2]**

Total Marks .... / 9

Pathogens and Disease

1 **a)** Draw **one** line from each disease to the type of microorganism that causes it.

Disease	Type of Microorganism
rose black spot	protozoan
salmonella	fungus
measles	virus
malaria	bacterium

[3]

b) Complete the sentences using the words from the list.

host   *Anopheles*   *Plasmodium*   **protozoan**   **red**   **vectors**

Mosquitoes are _____ because they carry malaria.

The mosquito's blood carries the malaria pathogen, which is called _____ .

The mosquito is a parasite because it feeds on a living organism,

called the _____ .   [3]

2 Human Immunodeficiency Virus (HIV) is a pathogen that can cause AIDS.

a) Give **two** ways that HIV can be passed on from one person to another.

_____

_____   [2]

b) AIDS stands for Acquired Immunodeficiency Syndrome.

Why is it called 'immunodeficiency'?

_____

_____   [2]

3 A gardener discovers that one of his rose bushes has black spot disease.
A friend advises him to burn any infected leaves rather than put them in the compost heap.

Why does she suggest this?

_____

_____   [2]

Total Marks _____ / 12

Human Defences Against Disease

1   **Figure 1** shows some of the main ways that pathogens can enter the body.

Figure 1

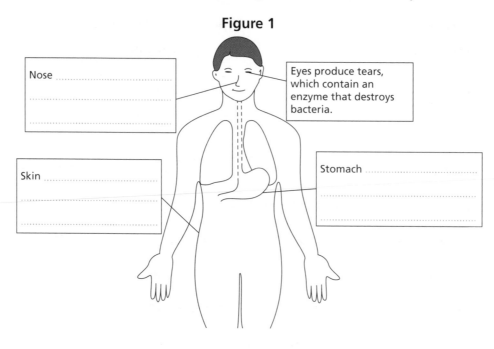

Nose ....
....
....

Eyes produce tears, which contain an enzyme that destroys bacteria.

Skin ....
....
....

Stomach ....
....
....

The diagram shows how the eyes work to stop pathogens entering the body.

Complete the diagram to show how the other **three** areas labelled prevent infection.   **[3]**

2   The four diagrams in **Figure 2** show stages in the immune response to a pathogen.

Figure 2

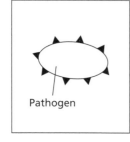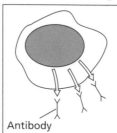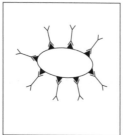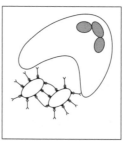

Pathogen

Antibody

Use the diagrams in **Figure 2** to explain how the immune system deals with a pathogen.

....

....

....

....

....

   [4]

Total Marks .... / 7

Treating Diseases

1 This article is about the treatment of tuberculosis.

Antibiotic Could Beat TB

Scientists believe that they have found an antibiotic that could beat tuberculosis (TB).

TB is a disease of the respiratory system and is caused by a bacterium.

TB almost disappeared but is now killing more people. This is because the antibiotics that were used to treat TB are not effective now.

The new antibiotic is called linezolid and is being tested in a double-blind trial.

---

a) People who have TB can be treated with antibiotics, but HIV cannot be treated in this way.

Give the reason why.

_____ [2]

b) Explain why TB is now on the increase.

_____

_____ [2]

c) Describe **two** precautions that can be taken when using antibiotics to help prevent problems like this occurring.

_____

_____ [2]

d) The article says that the new drug is being tested in a double-blind trial.

Explain what is meant by a double-blind trial.

_____

_____

_____ [3]

Total Marks _____ / 9

Photosynthesis

1 Plants make glucose by photosynthesis.

a) Write down the word equation for this process.

.... **[2]**

b) i) Glucose is transported from the leaves to other parts of a plant.

Suggest **two** parts of a plant that the glucose might be taken to.

....

.... **[2]**

ii) The glucose can also be built-up into different substances.
These substances can then be used in many different ways.

Name **two** of these substances and explain how they are used in a plant.

....

....

....

.... **[4]**

2 The passage is about early investigations into plant growth.

In the seventeenth century, the biologist, Jan Baptiste van Helmont, grew a tree in a bucket of soil. He fed the tree with rainwater only.

In five years, the tree had grown but the amount of soil in the bucket did not decrease. Baptiste concluded that extra material in the tree must have come from the rainwater.

In the eighteenth century, Jean Senebier showed that plants that live in water photosynthesise faster when they are supplied with extra carbon dioxide dissolved in the water.

a) To what extent was Van Helmont's conclusion correct?

....

.... **[2]**

b) How did the work of Senebier show that Van Helmont's conclusion was not completely correct?

.... **[1]**

Total Marks .... / 11

Respiration and Exercise

1. When Joanne runs a race her muscles work hard.
Her muscles use aerobic respiration to release energy from glucose.

a) Complete the word equation for aerobic respiration. [3]

glucose + ... $\longrightarrow$ .... + ...

b) Joanne is training to run a marathon.

When she runs, her muscles start to make lactic acid and this passes into her blood.

The graph in **Figure 1** shows the lactic acid concentration in Joanne's blood during her first training run and during the first part of the race.

Figure 1

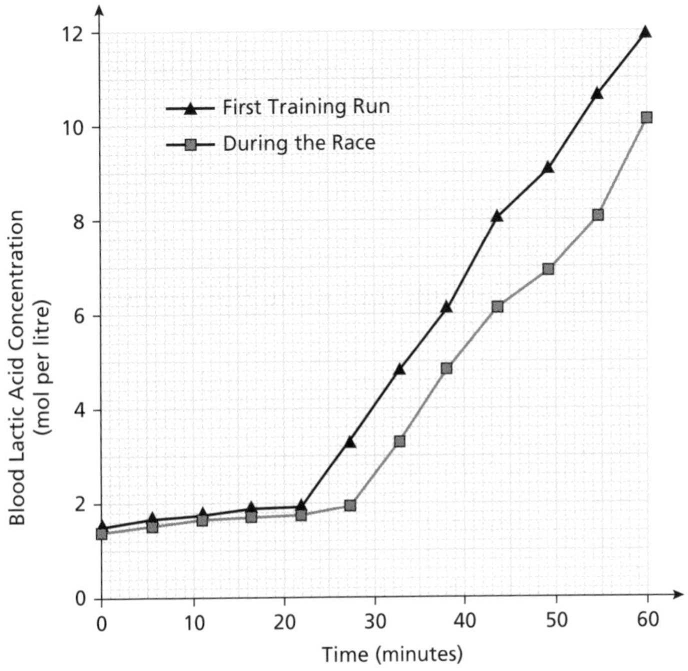

i) Use the graph to explain why Joanne can run more efficiently when she has trained for a race.

....

....

....

....

[3]

ii) Describe what happens to the lactic acid in the blood after Joanne stops running.

....

....

....

[3]

Total Marks .... / 9

Homeostasis and the Nervous System

1 **Figure 1** shows part of the nervous system.

The enlarged section shows a synapse. Transmitter molecules carry signals across the synapse.

Figure 1

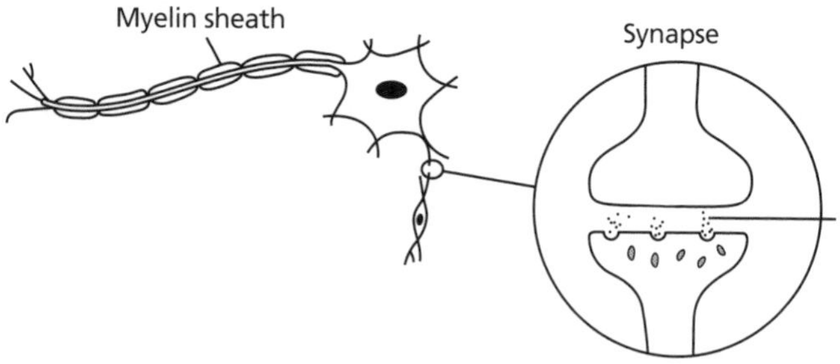

a) What are the functions of the myelin sheath (protective covering)?

[2]

b) Describe the function of the transmitter molecules at a synapse.

[3]

c) Many drugs or poisons work by affecting the action of synapses.

For parts **i)** and **ii)**, suggest how the different drugs or poisons described might work.

i) Some drugs stimulate the nervous system.
These drugs have a similar shape to neurotransmitter molecules.

[2]

ii) Neurotransmitter molecules are usually broken down by enzymes after they have stimulated the next neurone.
Some insect poisons block these enzymes.

[2]

Total Marks _____ / 9

Hormones and Homeostasis

1 This is an extract from an article published in a magazine.

> **Ricky's Story**
>
> Ricky is 16 years old and found out that he had diabetes when he was 10.
>
> 'At first it was a big change because I had to learn to do my injections and not to eat all the sweet things I used to eat. Even with my diabetes, I still play sport regularly and I'm hoping to become a PE teacher.'

a) What type of diabetes does Ricky have?   Answer: .... **[1]**

b) What would Ricky be injecting into his body?   Answer: .... **[1]**

c) Suggest what problems there would be if Ricky did not inject himself.

....

.... **[2]**

d) Ricky's uncle Gary also has diabetes.
   Gary is 50 years old and has only just developed diabetes.

   i) What type of diabetes does Gary have?   Answer: .... **[1]**

   ii) How would the treatment for Gary's diabetes differ from Ricky's treatment?

   ....

   .... **[2]**

2 Alcohol causes the pituitary gland to release too little ADH.

Explain what effects this might have on the body.

....

....

.... **[3]**

Total Marks .... / 10

Hormones and Reproduction

1 The graph in **Figure 1** shows the levels of two hormones in a woman's body at different stages of the menstrual cycle.

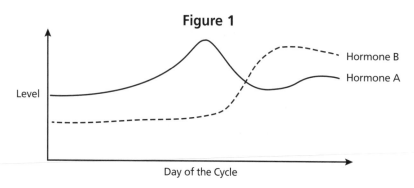

Figure 1

a) Both hormones are produced by the ovaries. Complete the following table.

	Name of Hormone	One Function of Hormone
Hormone A		
Hormone B		

[4]

b) Mark on the graph the approximate time of the following events:

i) Mark the time of ovulation with an **X**. [1]

ii) Mark the time of menstruation with a **Y**. [1]

c) Two other hormones are involved in the menstrual cycle.

i) The levels of FSH are high in the first half of the cycle. Explain why.

_____

_____ [2]

ii) The levels of LH are highest in the middle of the cycle. Explain why.

_____ [1]

2 Explain how the hormones in the contraceptive pill can prevent pregnancy occurring.

_____

_____

_____ [3]

Total Marks _____ / 12

Sexual and Asexual Reproduction

1 **Figure 1** shows a strawberry plant reproducing both sexually and asexually.

Figure 1

Plantlet

a) Describe how the plant reproduces asexually.

....

.... **[2]**

b) Describe how the plant reproduces sexually.

....

.... **[2]**

2 **Table 1** shows the number of chromosomes in different cells in a cat.
It also shows the type of cell division that produces these cells.

Complete the table.

Table 1

Cell Type	Number of Chromosomes	Type of Cell Division Used to Produce the Cell
Egg cell	19	
Eye cell		
Sperm cell		meiosis
Leg cell		mitosis

[4]

Total Marks .... / 8

Patterns of Inheritance

1 Gaucher disease (GD) is a genetic condition caused by a recessive allele.
People with Gaucher disease cannot make a particular enzyme.
The enzyme is needed to stop fat from building up in many organs, so the disease can be fatal.
Figure 1 shows part of a family tree showing some individuals that have Gaucher disease.

Figure 1

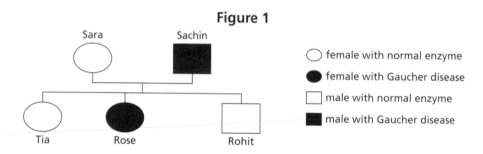

○ female with normal enzyme

● female with Gaucher disease

□ male with normal enzyme

■ male with Gaucher disease

a) Write down the names of all the people who are:

    i) **Definitely** homozygous for this gene.   Answer: .... **[1]**

    ii) **Definitely** heterozygous for this gene.   Answer: .... **[1]**

b) **Table 1** gives information about the different family members. Complete the table.

Table 1

Family Member	Possible Genotype	Phenotype
Sachin		
Tia	Gg	

[3]

c) Sara and Sachin are expecting another child.

   Work out the probability that the child will have Gaucher disease. Use a table like **Table 1** to explain your thinking.

                                      Probability: .... **[4]**

d) A pregnant woman can have her unborn baby tested for Gaucher disease.

   Suggest why she might decide **not** to have her baby tested even if it may have Gaucher disease.

....

....

[2]

Total Marks .... / 11

Variation and Evolution

1   Bill and Ben are identical twins.
    This means that they have inherited the same genes from their parents.

    Write each of the characteristics from **Figure 1** in the correct column of the table.

Figure 1

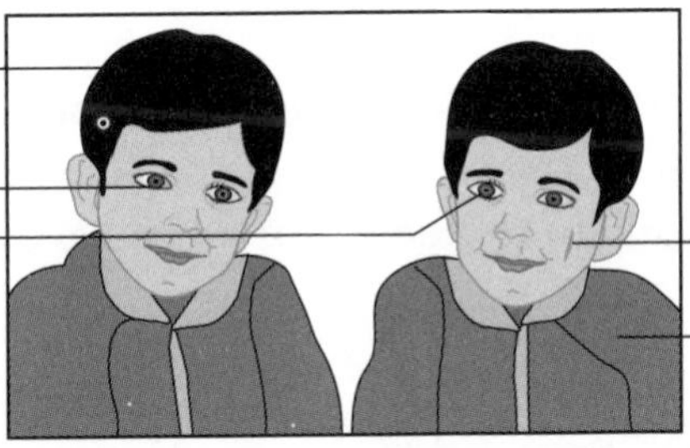

Bill is 160cm tall

Bill and Ben have brown eyes

Ben has a scar

Ben's body mass is 60kg

Controlled by Their Genes	Caused by the Environment	Controlled by their Genes and the Environment

[4]

2   **Figure 2** shows two forms of a moth that rests on trees.
    These moths are eaten by birds.

    Figure 2

a) When large amounts of coal were burned the air was heavily polluted with soot.

    Suggest what could happen to the colour of tree trunks in a heavily polluted area.

    _____ [1]

b) A survey of dark coloured and light coloured moths was carried out in a polluted area.
    An equal number of light and dark moths were collected, marked and released.
    Several days later, moths were recaptured.

    Suggest why more marked dark moths were recaptured than marked light moths.

    _____

    _____

    _____ [2]

Total Marks _____ / 7

Manipulating Genes

1. The following sentences are about processes that involve manipulating genes

Complete the sentences with words from the list.

cloning   **genetic engineering**   **selective breeding**   **tissue culture**

Farmers manipulate the genes of animals, by carefully selecting which animals mate.

This is called ....

It is now possible to make bacteria produce human insulin by the process of .... **[2]**

2. **Figure 1** shows some of the stages in the production of human insulin by bacteria.
The stages are not in the correct order.

Figure 1

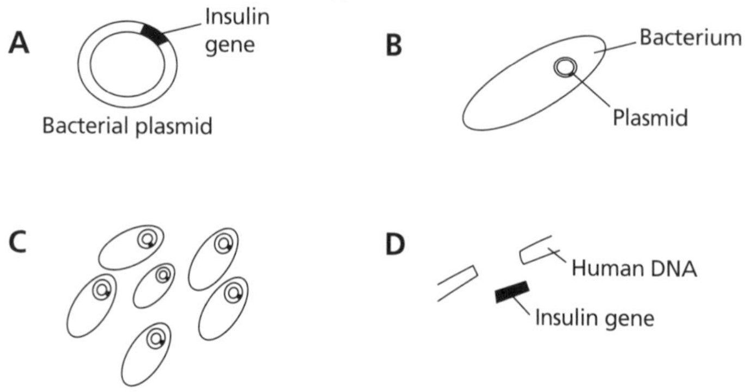

A Bacterial plasmid — Insulin gene

B Bacterium — Plasmid

C

D Human DNA — Insulin gene

a) Write down the order in which the steps should take place.
The first one has been done for you.

D   ....   **[2]**

b) Which of the statements about this method of making insulin are **incorrect**?
Put a cross next to **two** statements.

The process uses hormones to cut DNA. ☐

Vectors take pieces of DNA and insert them into other cells. ☐

The insulin molecules produced are a mixture of human and
bacterial insulin. ☐

The bacteria can be grown on cheap waste products. ☐   **[2]**

Total Marks .... / 6

Classification

1 **a)** What is meant by the term 'species'?

_____

_____ **[2]**

b) **Table 1** shows the Latin names of some different cats.

Table 1

  i) Two of the cats are more closely related than the others.

     Write down the common names of these two cats.

Common Name	Latin Name
bobcat	*Felis rufus*
cheetah	*Acinonyx jubatus*
lion	*Panthera leo*
ocelot	*Felis pardalis*

_____ **[2]**

  ii) Explain your answer to part **i)** above.

_____

_____ **[2]**

2 **a)** **Figure 1** shows an evolutionary tree, which includes an extinct animal called Archaeopteryx.
When Archaeopteryx became extinct humans did not exist.

Figure 1

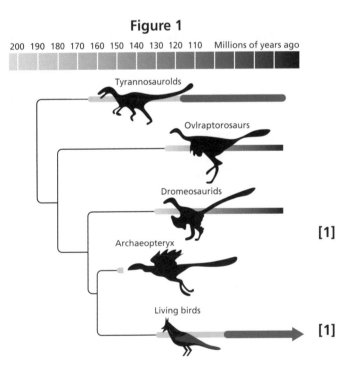

  i) How many years ago did Archaeopteryx and birds share a common ancestor?

     Answer: _____ **[1]**

  ii) Which organism shown on the tree existed for the shortest period of time?

     Answer: _____ **[1]**

b) Suggest **two** factors that may have led to its extinction.

_____

_____ **[2]**

Total Marks _____ / 10

Ecosystems

1 Camels live in desert areas of Africa.

a) Explain how the following special adaptations help the camel survive in the desert.

Figure 1

i) Webs of skin between their toes.

_____ **[1]**

ii) A store of fat in a hump on the top of their body.

_____ **[1]**

b) About one hundred years ago, camels were introduced into Australia.

Complete the sentences using words from the list.

abiotic   biotic   community   population   habitat   competed   reproduced

The camels that were introduced formed a large interbreeding _____ .

The area that they were introduced into was a new _____ for them and they were well adapted to live there.

The _____ made up of other plants and animals living there started to suffer.

The camels _____ with cattle for _____ factors such as food.   **[5]**

c) Scientists want to estimate how many organisms there are in an area of Australia.

i) Describe how they could estimate the number of cacti growing in that area.

_____

_____

_____

_____

_____   **[6]**

ii) Why is it harder to estimate the number of camels in the area?

_____

_____   **[2]**

Total Marks _____ / 15

Cycles and Feeding Relationships

1 **Figure 1** shows part of the carbon cycle.

Complete the diagram by writing the names of the correct processes in the empty boxes.

Figure 1

```
                          ┌──────────────────────────┐
          ┌──────────────▶│  Carbon dioxide in air   │◀──────┬──────────┐
          │               └──────────────────────────┘       │          │
          │                        ▲                          │          │
┌─────────────┐      ┌─────────────────────────┐    ┌──────────────┐   ┌──────────────┐
│ A           │      │ B                       │    │ C            │   │ D            │
└─────────────┘      └─────────────────────────┘    └──────────────┘   └──────────────┘
       │                   ▲          ▲                    ▲                  ▲
       │                   │          │          Carbon in waste             │
       │                   │          │          and dead remains            │
       │                   │          │               ▲    ▲          Carbon in
       ▼                   │          │               │    │          fossil fuels
Carbon in sugars, cellulose ──▶ ┌──────────┐     Carbon stored in              ▲
and starches in green plants    │ E        │ ──▶ animals bodies                │
       │                        └──────────┘               │                   │
       │                              ▼                     │                   │
       └──────────────────────▶ ┌──────────┐       Dead animals                │
                                │  Death   │ ────▶ and plants ─────────────────┘
                                └──────────┘
```

[5]

2 The graph in **Figure 2** shows how the number of rabbits and foxes in one square kilometre of woodland varies over a 40-month period.

Figure 2

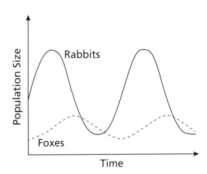

a) What name is given to this type of graph?

Answer: _____ **[1]**

b) Explain why the fox numbers change throughout the 40 months.

_____

_____

_____ **[3]**

Total Marks _____ / 9

Disrupting Ecosystems

1 **Figure 1** shows the levels of carbon dioxide in the air above a small island in the Pacific Ocean. Scientists measured the carbon dioxide here to see if levels in the Earth's atmosphere are increasing.

Figure 1

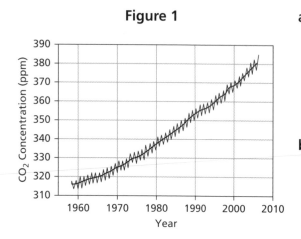

a) Suggest why scientists measured the levels at this remote island.

....

.... **[1]**

b) Levels of carbon dioxide go up slightly each winter and down each summer.

Suggest why this might be.

....

....

.... **[2]**

c) Explain how increasing carbon dioxide levels could lead to global warming.

....

....

.... **[3]**

2 When acid rain falls into a lake, the water becomes more acidic.
This may change the types of animals and plants that can live there.
Table 1 shows a number of aquatic animals and the range of pH in which they are found.

Table 1

Name of Animal	pH 4.0	pH 4.5	pH 5.0	pH 5.5	pH 6.0	pH 6.5
perch		✓	✓	✓	✓	✓
frog	✓	✓	✓	✓	✓	✓
clam					✓	✓
crayfish				✓	✓	✓

a) Which animal can survive the most acidic conditions?

Answer: .... **[1]**

b) Explain how acid rain is caused.

....

....

....

.... **[3]**

Total Marks .... / 10

Atoms, Elements, Compounds and Mixtures

1 A student heated a piece of copper in air.
The word equation for the reaction is:

copper + oxygen → copper(II) oxide

a) Which equation correctly represents the reaction?
Tick **one** box.

$Cu + O → CuO$ ☐

$2Cu + O_2 → 2CuO$ ☐

$2Cu + 2O → 2CuO$ ☐

$Cu_2 + O_2 → 2CuO$ ☐ **[1]**

b) The mass of the copper before heating was 15.6g.
The mass of copper oxide produced was 18.9g.

Calculate the mass of oxygen that was used in the reaction.

Answer: _____ g **[1]**

c) What was the resolution of the balance the student used to measure the masses?

Answer: _____ g **[1]**

2 A student was asked to produce pure water from salt solution.

a) Describe how they would do this using the equipment in **Figure 1**.

_____

_____

_____

_____ **[3]**

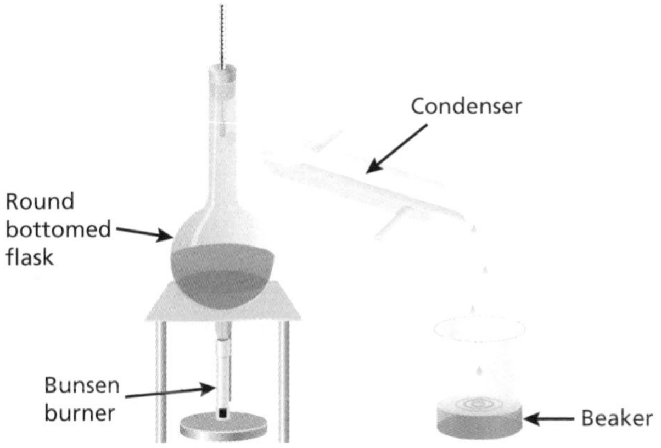

b) State **one** hazard when carrying out this procedure.

_____ **[1]**

Figure 1

Condenser

Round bottomed flask

Bunsen burner

Beaker

Total Marks _____ / 7

Atoms and the Periodic Table

1 Early models of atoms showed them as tiny spheres that could not be divided into simpler substances.

Figure 1

In 1897, Thomson discovered that atoms contained small negatively charged particles. He proposed a new model shown in **Figure 1**.

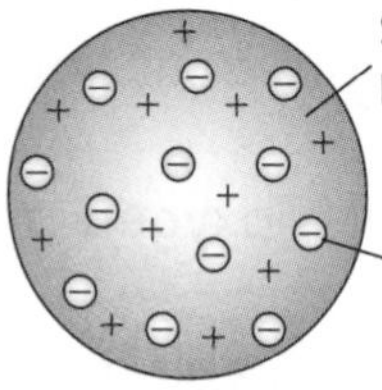

a) Why did the model of the atom have to change?

[2]

b) Name the particle labelled **X**.

Answer: [1]

2 In 1909, Geiger and Marsden bombarded a thin sheet of gold with positively charged alpha particles. They found that most passed through, but some were deflected back.

a) Explain why this result was unexpected.

[1]

b) Explain why they would have repeated their experiment several times.

[2]

c) Describe the new model of the atom that resulted from this experiment.

[3]

Total Marks _____ / 9

The Periodic Table

1 An unknown element has the electronic configuration 2,8,8,7.

Where would it be found in the periodic table shown in **Figure 1**?

Tick **one** box.

Figure 1

A ☐   B ☐

C ☐   D ☐

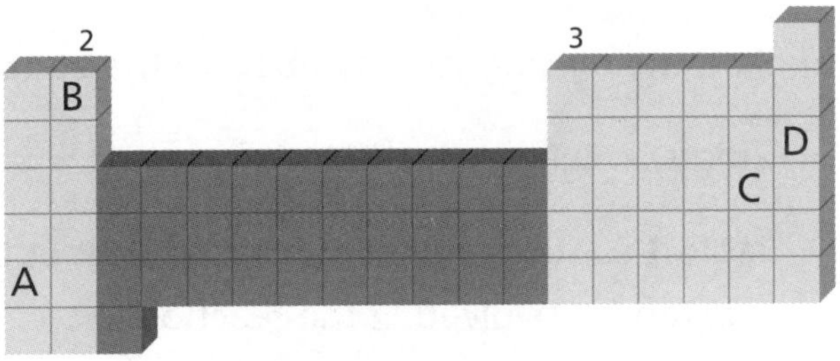

[1]

2 **Table 1** shows some data on the physical properties of elements.

Table 1

Physical Property	Element W	Element X	Element Y	Element Z
Melting Point (°C)	−38.82	−189.34	180.5	1538
Density (g/cm³)	13.53	1.40	0.53	7.87
Conductor of Electricity?	Yes	No	Yes	Yes

a) Which element, **W, X, Y** or **Z**, is a non-metal?
   Give a reason for your answer.

   ....

   .... [2]

b) Which element, **W, X, Y** or **Z**, is mercury?
   Give a reason for your answer.

   ....

   ....

   .... [3]

c) One of the elements is the Group 1 metal lithium.
   Complete the word equation to show the reaction of lithium with water.

   lithium + water → lithium .... + ... [2]

Total Marks .... / 8

States of Matter

1 A student burns magnesium.

It reacts with oxygen in the air to form a white powder called magnesium oxide.

a) Add the missing state symbols to the equation for the reaction.

$2Mg(s) + O_2(\text{\_\_\_\_}) \rightarrow 2MgO(\text{\_\_\_\_})$

[2]

Table 1 contains some information about the melting and boiling points of the substances involved in the reaction.

Table 1

	Magnesium	Oxygen	Magnesium Oxide
Melting Point (°C)	650	−219	2830
Boiling Point (°C)	1091	−183	3600

b) State the temperature at which magnesium would change from a solid into a liquid.

Answer: _____ °C   [1]

c) State the temperature that oxygen gas would have to be cooled to in order for it to condense.

Answer: _____ °C   [1]

d) Explain, in terms of bonding, the difference in boiling points between oxygen and magnesium oxide.

_____

_____

_____

_____

_____

_____

_____

_____

_____

[6]

Total Marks _____ / 10

Ionic Compounds

1 A student investigated the properties of some different compounds.

   a) Which of the following compounds contain ionic bonds?
      Tick **two** boxes.

      water (H_2O) ☐   glucose ($C_6H_{12}O_6$) ☐

      calcium chloride ($CaCl_2$) ☐   sodium carbonate (Na_2CO_3) ☐

      hydrogen chloride (HCl) ☐   **[2]**

   b) The student discovered that the ionic compounds would not conduct electricity when solid but they would when dissolved in water.

      Explain why.

      ....

      ....

      .... **[3]**

2 **Figure 1** shows the outer electrons in an atom of the Group 2 element magnesium and in an atom of the Group 7 element bromine.
Magnesium forms an ionic compound with bromine.

Describe what happens when **one** atom of magnesium reacts with **two** atoms of bromine.
Give your answer in terms of electron transfer.
Give the formulae of the ions formed.

Figure 1

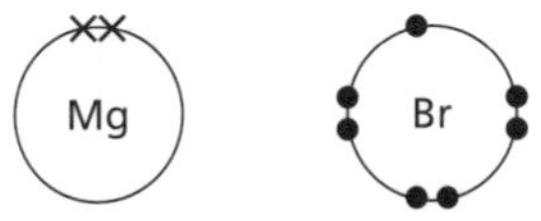

....

....

....

....

....

.... **[5]**

Total Marks .... / 10

Covalent Compounds

1 Fluorine and bromine are elements found in Group 7 of the periodic table.
At room temperature fluorine is a gas and bromine is a liquid.

Why, at room temperature, is fluorine a gas and bromine a liquid?
Tick **one** box.

The covalent bonds between bromine are stronger. ☐

Bromine has a giant covalent structure and fluorine is a simple molecule. ☐

The forces between bromine molecules are stronger. ☐

Fluorine contains fewer molecules than bromine. ☐ **[1]**

2 Bromine reacts with hydrogen to form hydrogen bromide.

Figure 1

a) Complete the dot and cross diagram in **Figure 1** to show the covalent bonding in a molecule of hydrogen bromide.

Show the outer shell electrons only.

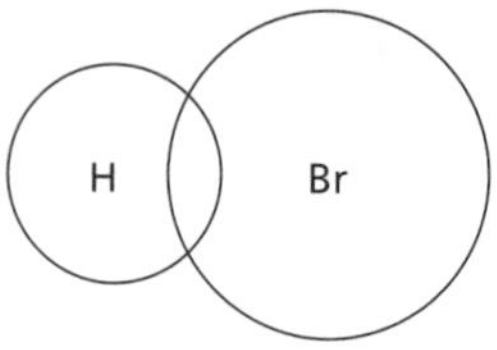

[2]

b) State the formula for a molecule of hydrogen bromide.

Answer: _____ **[1]**

3 Diamond has a giant covalent structure made up of carbon atoms.
It has a high melting point and does not conduct electricity when molten.

Draw **one** line from each property to the explanation of that property.

Property	Explanation of Property
	Strong covalent bonds between many carbon atoms
High melting point	Atoms are free to move
	Weak bonds between carbon atoms
Does not conduct electricity when molten	There are no charged particles that are free to move
	Carbon atoms are in a regular arrangement

[2]

Total Marks _____ / 6

Metals and Special Materials

1 **Figure 1** shows the bonding in the metal copper.

Figure 1

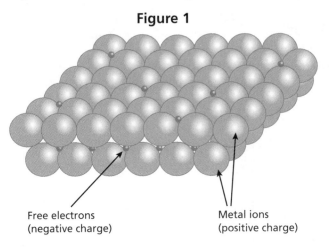

Free electrons
(negative charge)

Metal ions
(positive charge)

Copper is useful as a material for making saucepans.

This is because it has the following properties:

- high melting point
- good thermal conductor
- malleable (can be easily shaped).

Explain, in terms of its metallic bonding, why copper has these properties.

[6]

Total Marks _____ / 6

Conservation of Mass

1 A student was asked to carry out the reaction shown in the word equation:

zinc + hydrochloric acid → zinc chloride + hydrogen

a) What is the ionic equation for the reaction?
   Tick **one** box.

   $Zn(s) + 2HCl(aq) \rightarrow ZnCl_2(aq) + H_2(g)$ ☐

   $Zn^{2+}(aq) + 2Cl^-(aq) \rightarrow ZnCl_2(aq)$ ☐

   $Zn(s) + 2H^+(aq) \rightarrow Zn^{2+}(aq) + H_2(g)$ ☐

   $Zn(s) + 2Cl^-(aq) \rightarrow ZnCl_2(aq)$ ☐ [1]

b) The student plans to use the equipment in **Figure 1**.
   The teacher tells the student not to use the bung.

   Explain why.

   _____

   _____

   _____

   _____ [2]

Figure 1

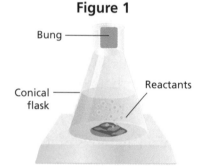

Bung

Conical flask

Reactants

c) The student carries out the reaction on a balance, as shown in **Figure 2**.

Figure 2

Explain what happens to the mass reading during the reaction.

_____

_____

_____

_____

_____ [3]

Total Marks _____ / 6

Amount of Substance

1 **Figure 1** shows a cube of pure silver with a side length of 2cm.

Figure 1

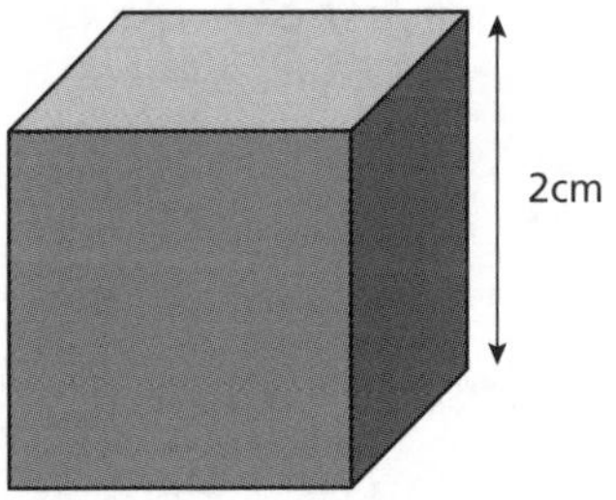

2cm

a) Calculate the volume of the cube.

Answer: _____ cm³ **[1]**

b) The density of silver is 10.49g/cm³.

Use the formula **density** = $\dfrac{\textbf{mass}}{\textbf{volume}}$ to calculate the mass of the silver cube to 2 decimal places.

Answer: _____ g **[2]**

c) Calculate how many moles of silver are in the cube to 2 decimal places.
(Relative atomic mass, A_r, of silver = 108)

Answer: _____ moles **[2]**

d) There are 6.02×10^{23} atoms in one mole of silver.

Which of the following also contains 6.02×10^{23} atoms?
(Relative atomic masses, A_r: C = 12, O = 16)
Tick **two** boxes.

1 mole of oxygen (O_2)   ☐

0.5 mole of oxygen (O_2)   ☐

12g of carbon (C)   ☐

6g of carbon (C)   ☐   **[2]**

Total Marks _____ / 7

Reactivity of Metals

1 This question is about how metals are extracted from their ores.
Figure 1 shows the reactivity series of metals.

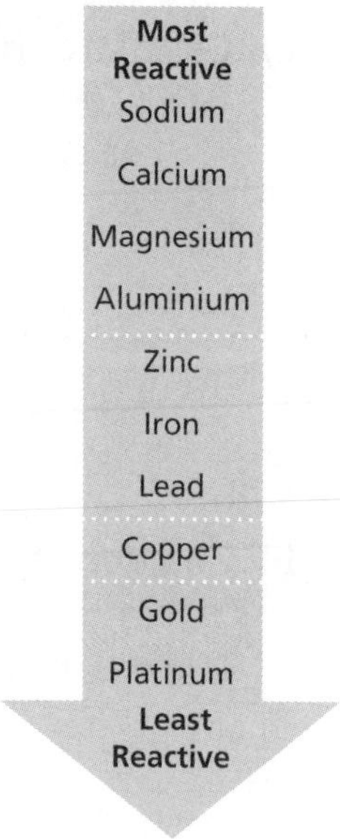

a) Zinc can be found naturally as the compound zinc(II) oxide.

Use the information in **Figure 1** to decide which of these metals can be
used to displace zinc from zinc(II) oxide.
Tick **two** boxes.

magnesium ☐   iron ☐

copper ☐   sodium ☐   **[2]**

Figure 1

Most
Reactive
Sodium

Calcium

Magnesium

Aluminium

Zinc

Iron

Lead

Copper

Gold

Platinum
Least
Reactive

b) In reality, carbon is used to extract zinc from zinc(II) oxide.
The equation for the reaction is:

zinc(II) oxide + carbon → zinc + carbon monoxide

i) State the substance that is oxidised during the reaction.

Answer: _____ **[1]**

ii) State the substance that is reduced during the reaction.

Answer: _____ **[1]**

2 A student reacted zinc with a solution of copper chloride:

$Zn + CuCl_2 \rightarrow ZnCl_2 + Cu$

Draw **one** line from each change that takes place in this reaction to the correct
half equation.

Change

| Oxidation |

| Reduction |

Half Equation

| $Cu \rightarrow Cu^{2+} + 2e^-$ |

| $Zn \rightarrow Zn^{2+} + 2e^-$ |

| $Zn^{2+} + 2e^- \rightarrow Zn$ |

| $Zn^{2-} \rightarrow 2e^- + Zn$ |

| $Cu^{2+} + 2e^- \rightarrow Cu$ |

[2]

Total Marks _____ / 6

The pH Scale and Salts

1 Universal indicator was added to a sample of an unknown solution.
The universal indicator turned orange.

What is the pH value of the solution?
Tick **one** box.

1 ☐   7 ☐

4 ☐   10 ☐   [1]

2 A student was asked to prepare pure, dry crystals of a soluble salt using insoluble copper(II) oxide and hydrochloric acid.
First, the student measured out 25cm³ of acid into a conical flask.

a) Describe how the student should complete the preparation of the salt.

....

....

....

....

....

....

....

....   [4]

b) State **one** safety precaution that the student should take during the preparation.

....

....   [1]

c) Name the salt produced.

Answer: ....   [1]

Total Marks .... / 7

Electrolysis

1. Aluminium is extracted from its ore, aluminium oxide, using electrolysis, as shown in **Figure 1**.

Figure 1

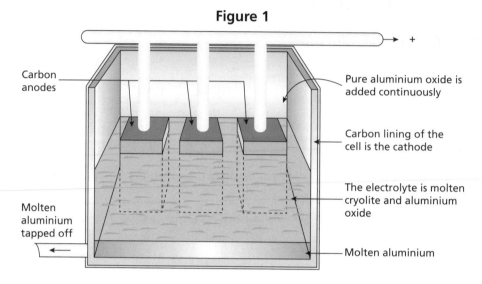

Carbon anodes

Pure aluminium oxide is added continuously

Carbon lining of the cell is the cathode

The electrolyte is molten cryolite and aluminium oxide

Molten aluminium tapped off

Molten aluminium

a) Explain why aluminium cannot be extracted from aluminium oxide using reduction with carbon.

....

....

.... **[1]**

b) Balance the half equation for the reaction that occurs at the carbon anodes.

.... O^{2-} → .... $e^- + O_2$ **[2]**

c) Explain why the extraction of aluminium from aluminium oxide is an expensive process and describe the methods used by manufacturers to reduce the cost.

....

....

....

....

....

....

....

.... **[3]**

Total Marks .... / 6

Exothermic and Endothermic Reactions

1 A student carried out an investigation into how the reactivity of metals affects how exothermic or endothermic their reaction with dilute hydrochloric acid is.

The student used the apparatus shown in **Figure 1**.

The student's results are shown in **Table 1**.

Figure 1

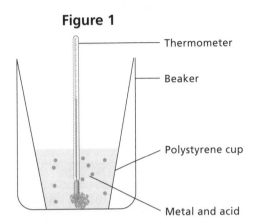

Table 1

Metal	Temperature at Start (°C)	Highest Temperature Reached (°C)
Zinc	21.0	30.1
Copper	21.2	21.3
Magnesium	21.4	82.6
Iron	21.4	26.0

a) Describe the function of the polystyrene cup.

....

....

.... **[2]**

b) State **two** control variables the student should have used in this investigation.

....

.... **[2]**

c) The order of reactivity of these metals from highest to lowest is: magnesium, zinc, iron, copper.

Write a conclusion that answers the original question that the student was investigating.

....

....

.... **[2]**

Total Marks .... / 6

Measuring Energy Changes

1 **Figure 1** shows the displayed formulae for the decomposition of hydrogen bromide.

Figure 1

H–Br

————→ H–H + Br–Br

H–Br

The bond enthalpies for the reaction are shown in **Table 1**.

Table 1

	H–Br	H–H	Br–Br
Energy (kJ/mol)	366	436	193

a) Calculate the energy needed to break the bonds in the reactants.

Answer: .... kJ/mol   **[2]**

b) Calculate the energy released when the bonds in the products are formed.

Answer: .... kJ/mol   **[2]**

c) Is this reaction **endothermic** or **exothermic**?

Explain your answer.

....

....

.... **[2]**

Total Marks .... / 6

Rate of Reaction

1 When hydrochloric acid reacts with sodium thiosulfate one of the products is sulfur, which is insoluble.

A student carried out an investigation into the following hypothesis:

As the concentration of acid increases, the rate of the reaction will increase.

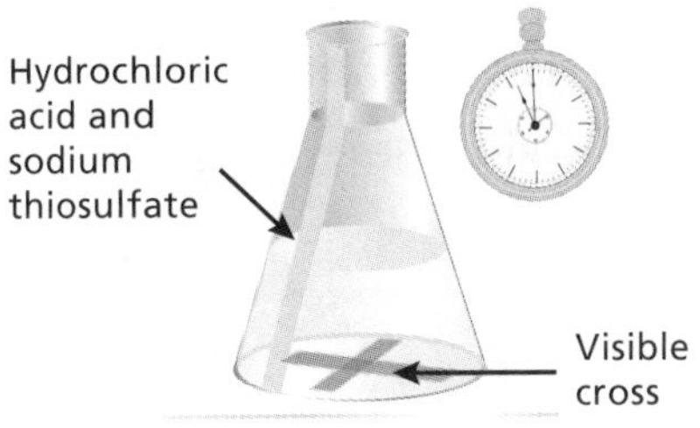

The student added sodium thiosulfate to dilute hydrochloric acid of concentration $0.25\,mol/dm^3$ in a conical flask.
The conical flask was placed on a cross, as shown in **Figure 1**.
The student timed how long it took before they could no longer see the cross.

Figure 1

Hydrochloric acid and sodium thiosulfate

Visible cross

a) Describe what the student should do next in order to investigate the hypothesis. You should include a suitable range for the independent variable in your answer.

....

....

....

.... **[2]**

b) State **one** safety precaution the student should take.

.... **[1]**

c) Use collision theory to write a prediction for what will happen if the hypothesis is correct.

....

....

....

....

....

.... **[4]**

Total Marks .... / 7

Reversible Reactions

1 This question is about the reaction between ammonia and hydrogen chloride.
The equation for the reaction is:

$$NH_3(g) + HCl(g) \rightleftharpoons NH_4Cl(s)$$

The reaction can be carried out using the apparatus shown in **Figure 1**.

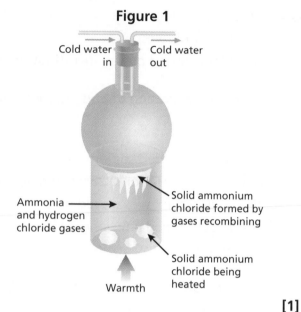

Figure 1

Cold water in

Cold water out

Solid ammonium chloride formed by gases recombining

Ammonia and hydrogen chloride gases

Solid ammonium chloride being heated

Warmth

a) What is meant by the symbol $\rightleftharpoons$ in the equation?

Answer: .... [1]

b) Which reaction, the **forward reaction** or the **reverse reaction**, is exothermic?
Give a reason for your answer.

....

....

.... [3]

c) The reaction taking place in **Figure 1** is at equilibrium.

What does this mean?
Tick **two** boxes.

Only ammonium chloride is being produced. ☐

The forward and reverse reactions are taking place at the same rate. ☐

The amounts of reactants and products are constant. ☐

The reaction has completed. ☐ [2]

d) Predict what will happen if more ammonia is added to the mixture.

....

....

.... [2]

Total Marks .... / 8

Alkanes

1 Fractional distillation is used to separate crude oil into useful mixtures called fractions.
It takes place in a column, as shown in **Figure 1**.

a) The molecules found in crude oil are mostly hydrocarbons.

What is a hydrocarbon?

....

....

....

.... [2]

Figure 1

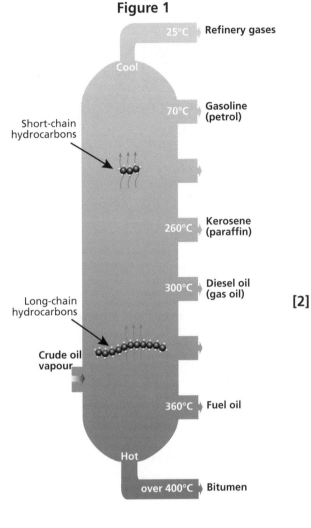

b) This is what takes place inside the column:

1. Crude oil is heated until it forms a vapour.
2. The vapour rises up the tower.

Describe what happens next in order for the fractions to form.

....

....

....

....

....

.... [2]

2 Methane (CH_4) is the name of a hydrocarbon found in refinery gas.

Complete the balanced symbol equation for the complete combustion of methane.

$CH_4 + 2$ .... $\rightarrow CO_2 +$ .... H_2O [2]

Cracking Hydrocarbons

1 Hydrocarbon molecules can be cracked to form molecules with shorter chains.
The equation shows the cracking of $C_{20}H_{42}$.

$$C_{20}H_{42} \rightarrow C_8H_{18} + \boxed{} C_5H_{10} + C_2H_4$$

a) What number needs to go in the box to balance the equation?
Tick **one** box.

1 ☐

2 ☐

3 ☐

4 ☐

[1]

b) Which products are alkanes?
Tick **one** box.

C_8H_{18} and C_2H_4 ☐

C_5H_{10} and C_2H_4 ☐

C_8H_{18} only ☐

C_8H_{18}, C_5H_{10} and C_2H_4 ☐

[1]

2 A student is given a test tube of colourless liquid hydrocarbon.

Describe a test the student could carry out to find out if the liquid is an alkane or an alkene.
You should state the chemical used for the test and the results of the test for both an alkane and an alkene.

[3]

Total Marks _____ / 5

Chemical Analysis

1 A student was asked to investigate five different inks using chromatography.

Figure 1 shows the apparatus they needed to use.

The student set up the apparatus correctly and left it for several minutes.

Figure 2 shows their results.

Figure 1

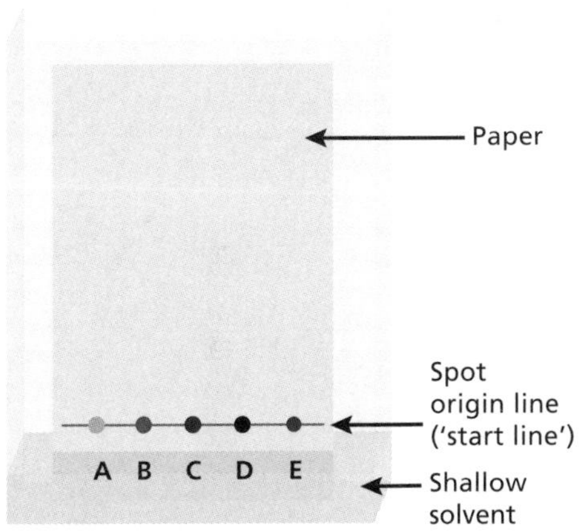

Paper

Spot origin line ('start line')

A B C D E

Shallow solvent

Figure 2

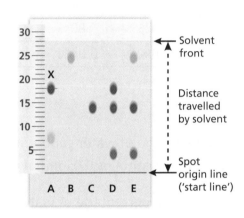

Solvent front

Distance travelled by solvent

Spot origin line ('start line')

A B C D E

a) Which part of the apparatus is the stationary phase?

Answer: _____ **[1]**

b) The student was not sure whether to use a pen or pencil to draw the origin line.

State which they should choose. Give a reason for your answer.

_____

_____ **[2]**

c) Which **two** inks are pure? Give a reason for your answer.

_____

_____ **[2]**

d) Calculate the R_f value of **X**.
Give your answer to 2 decimal places.

Answer: _____ **[3]**

Total Marks _____ / 8

The Earth's Atmosphere

1 **Table 1** shows some of the gases that are in the Earth's atmosphere today.

Table 1

	Oxygen	Nitrogen	Carbon Dioxide
A	80%	20%	0.04%
B	0.04%	80%	20%
C	20%	80%	0.04%
D	20%	0.04%	80%

Which letter, **A**, **B**, **C** or **D**, represents the correct proportions?
Tick **one** box.

A ☐

B ☐

C ☐

D ☐   **[1]**

2 The percentage of water vapour in the atmosphere can vary.
4.2dm³ of air contains 0.05dm³ of water vapour.

Calculate the percentage of water vapour in the air.
Give your answer to 2 significant figures.

Answer: _____ %   **[2]**

3 One theory for how the Earth's early atmosphere was formed is that gases were released by volcanoes.
Scientists have recently proposed a new theory: that the early atmosphere was formed by comets hitting the Earth.

a) What would the scientists need for this new theory to be accepted?

_____

_____   **[2]**

b) Explain why we cannot be sure if either theory is correct.

_____

_____

_____

_____   **[3]**

Total Marks _____ / 8

Greenhouse Gases

1 Many scientists believe that human activities are causing the mean global temperature of the Earth to increase.

They think that this is because of an increase in the amounts of greenhouse gases in the atmosphere.

Water vapour is an example of a greenhouse gas.

a) Name **one** other greenhouse gas and state how human activity has increased its amount in the atmosphere.

....

....

.... **[2]**

In 2000, computer models were used to predict how the mean global temperature might change in the future.

Figure 1 shows the results. Each line shows the predictions made using a different model.

Figure 1

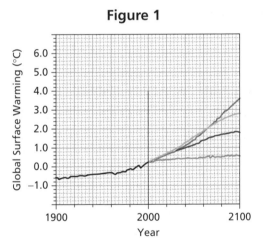

b) Calculate the range of global surface warming as predicted for 2100 by the different models.

From: .... To: .... **[2]**

c) Explain why it is difficult to produce models to predict future climate change.

....

....

.... **[2]**

Total Marks .... / 6

Earth's Resources

1 In the UK, potable water is produced from an unpolluted source of fresh water. Potable water contains low levels of dissolved substances.

a) What is potable water?

.... **[1]**

b) Explain why potable water cannot be called pure in the chemical sense.

....

....

.... **[2]**

The stages of producing potable water from fresh water are:

1. Pass the water through filter beds to remove any solids.
2. Sterilise the water to kill microorganisms.

c) Give **one** way of sterilising the water.

.... **[1]**

In other parts of the world, sea water is used as a source of potable water.
To produce potable water from sea water a process called distillation is used.

d) Explain why distillation is used.

....

.... **[1]**

e) Explain why producing potable water from sea water is more expensive than producing it from fresh water sources.

....

....

....

.... **[2]**

Total Marks .... / 7

Using Resources

1. Shopping bags can be made out of paper or plastic.

Table 1 is part of a LCA (life cycle assessment) comparing these two materials.

Table 1

	Paper Bag	Plastic Bag
Raw Materials	Wood pulp	Crude oil
Manufacture	• Wood pulp is added to water (up to 100 times the mass of the pulp) and mixed. • Clay, chalk or titanium oxide is added. • The mixture is squeezed and heated to remove the water.	• Crude oil is heated to 360°C during fractional distillation. • A fraction is cracked at 850°C to produce alkenes. • Polymerisation of alkenes at 150°C.
Transport	• Average mass = 55g • Seven trucks needed to transport two million bags.	• Average mass = 7g • One truck needed to transport two million bags.
Use During Lifetime	• Not normally reused.	• Can be reused many times.
Disposal at the End of Life	• Taken to landfill (biodegradable). • Can be incinerated (burned). • Can be recycled.	• Taken to landfill (not biodegradable). • Can be incinerated (burned). • Difficult to recycle.

Use the information in **Table 1** to compare the **advantages** and **disadvantages** of both types of bag.

[6]

Total Marks .... / 6

Forces – An Introduction

1 When a plane is in flight, the engines provide a thrust force that pushes the aircraft forwards. The wings provide a 'lift' force that acts upwards.

a) Name **two** other forces that act on the plane.

In each case state whether it is contact force or non-contact force.

[4]

b) The plane has a mass of 120 000kg. Calculate the weight of the plane (g = 10N/kg).

Weight = _____ N   [2]

c) The plane accelerates and ascends to a higher altitude.
During this time, the resultant forwards force is twice the size of the resultant upwards force.

Use this information to draw a scale
vector diagram.
Your diagram should show:
- The resultant forwards force.
- The resultant upwards force.
- A final resultant force that shows the
  combined effect of all the forces acting
  on the plane.

[3]

2 A student carries out an investigation into forces.
They use an air track, which suspends a glider vehicle on a cushion of air so that it can move smoothly without touching the ground.

a) What force is the air track designed to reduce?

Answer: _____   [1]

b) Use the idea of contact and non-contact forces to explain why the air track is effective at doing this.

[2]

Total Marks _____ / 12

Forces in Action

1 A student carries out an investigation involving springs.

The student suspends a spring from a rod.

A force is applied to the spring by hanging a mass from it.

a) Describe how the student could test if the spring is behaving elastically.

_____

_____ **[2]**

b) In a second investigation, the student takes a set of measurements for force and extension.
The results are shown in **Table 1** below.

Table 1

Force (N)	0.0	1.0	2.0	3.0	4.0	5.0	6.0
Extension (cm)	0.0	4.0		12.0	16.0	22.0	31.0

i) Add the missing value to the table. **[1]**

ii) Explain why you chose this value.

_____

_____ **[2]**

c) Complete the following sentences about the experiment in part **b)**.

The independent variable investigated was the _____ .

It had a range from _____ to _____ . **[3]**

d) The student repeats the experiment in part **b)**.

Give **two** reasons why repeating an experiment can improve the accuracy of the results.

_____

_____ **[2]**

Total Marks _____ / 10

Forces and Motion

1 A person takes their dog for a walk.

The graph in **Figure 1** shows how the distance from their home changes with time.

Figure 1

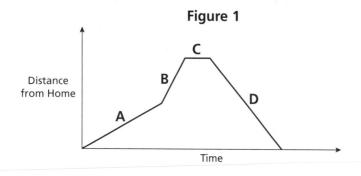

a) Work out the final displacement at the end of their journey.

Answer: _____ **[1]**

b) Which part of the graph, **A**, **B**, **C** or **D**, shows them walking at the fastest rate?

Answer: _____ **[1]**

c) Describe their motion during section **C**.   Answer: _____ **[1]**

d) Describe how the velocity of section **A** compares to the velocity of section **D**.

_____

_____

_____ **[3]**

2 Drivers on a racetrack enter a hairpin bend travelling east.

The tight bend forces them to slow down.

When they exit the bend, they are travelling west and speed back up again.

The bend is 180m long.

a) If it takes 6 seconds to travel around the bend, what is the average speed of the car around the bend?

Speed = _____ m/s **[2]**

b) A driver enters the bend at 50m/s and exits the bend at 40m/s.

i) Work out the change in speed.

Change in speed = _____ m/s **[1]**

ii) Work out the change in velocity.

Change in velocity = _____ m/s **[1]**

Total Marks _____ / 10

Forces and Acceleration

1 An experiment is carried out to investigate how changing the mass affects the acceleration of a system.

A trolley is placed on a bench and is made to accelerate by applying a constant force using hanging masses. Different masses were then added to the trolley.

Figure 1

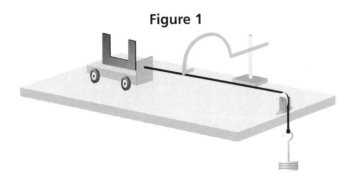

a) What is the independent variable?

Answer: _____ **[1]**

b) What is the control variable?

Answer: _____ **[1]**

c) What is the dependent variable?

Answer: _____ **[1]**

2 A boat accelerates at a constant rate in a straight line.
This causes the velocity to increase from 4.0m/s to 16.0m/s in 8.0s.

a) Calculate the acceleration.
Give the unit.

Answer: _____ **[3]**

b) A water skier being pulled by the boat has a mass of 68kg.

Use your answer to part **a)** to calculate the resultant force acting on the water skier whilst accelerating.

Answer: _____ **[2]**

3 The manufacturer of a car gives the following information in a brochure:
The mass of the car is 950kg.
The car will accelerate from 0 to 33m/s in 11s.

a) Calculate the acceleration of the car during the 11s.

Answer: _____ **[2]**

b) Calculate the force needed to produce this acceleration.

Answer: _____ **[2]**

Total Marks _____ / 12

Terminal Velocity and Momentum

1 The terminal velocity is the maximum velocity a falling object can reach.

Describe how a skydiver could increase their terminal velocity.

[2]

2 A Saturn 5 rocket, as used on the Apollo space missions, has F1 rocket engines.
Each engine burns 5000kg of fuel every second.
The exhaust gases leave the thrusters at 1340m/s.

a) Calculate the momentum of the exhaust gases.

Answer: .... kg m/s **[3]**

b) During the Apollo mission, five engines were used simultaneously.

Use your answer to part **a)** to calculate the momentum gained by the rocket in the first 10 seconds of travel (assuming there were no other resistive forces).

Answer: .... kg m/s **[3]**

c) The entire launch vehicle has a mass of 3000 000kg.

Use your answer to part **b)** to calculate the velocity after the first 10 seconds (assuming all other factors are unchanged).

Answer: .... m/s **[3]**

d) The rocket engines provide constant thrust.
However, after 100 seconds, the acceleration of the rocket is greater than the initial acceleration.

Give **two** reasons why this might be.
You must explain your answers.

[4]

Total Marks .... / 15

Stopping and Braking

1 The graph in **Figure 1** shows how the velocity of a car changes from the moment the driver sees an obstacle blocking the road.

Figure 1

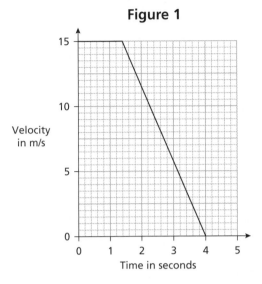

a) Work out the reaction time of the driver.

Answer: _____ [1]

b) Use the graph and your answer to part **a)** to calculate the thinking distance.

Answer: _____ [2]

c) Use the graph to work out the time it took for the car to stop from the moment the brakes were applied.

Answer: _____ [1]

d) The car and driver have a combined mass of 1300kg.

Work out the momentum of the car just before the brakes were applied.

Answer: _____ [2]

e) How would the time it took to stop the car be different it the road was wet or icy?

_____ [1]

f) The driver of the car was tired and had been drinking alcohol.

On the graph in **Figure 1**, sketch a second line to show how the graph would be different if the driver had been wide awake and fully alert. [3]

g) How would the graph look different if the vehicle had old / worn brakes and tyres?

_____

_____ [2]

Total Marks _____ / 12

Energy Stores and Transfers

1. An electric kettle is used to heat 2kg of water from 20°C to 100°C.

> **change in thermal energy = mass × specific heat capacity × temperature change**
>
> Specific heat capacity of water = 4200J/kg°C

a) All of the energy supplied to the kettle goes into the water.

Calculate the amount of electrical energy supplied to the kettle.

Answer: .... [3]

b) On a different occasion, the kettle is filled with 2.5kg of water but is switched on for the same amount of time.

Use your answer to part a) to calculate what temperature the kettle heats the water to in this period.

Answer: .... [3]

2. **Figure 1** shows a pendulum swinging backwards and forwards.
The pendulum is made from a 100g mass suspended by a light string.

Figure 1

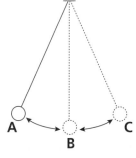

a) At position **C**, the mass is 5cm higher than at position **B**.

Calculate the difference in gravitational energy between positions **B** and **C**.
The gravitational field strength is = 10N/kg.

Answer: .... [3]

b) The difference in potential energy is the same as the amount of kinetic energy gained by the mass as it swings down from **C** to **B**.

Calculate the velocity of the mass at position **B**.

Answer: .... [3]

> **Total Marks** .... / 12

Energy Transfers and Resources

1 Complete the sentences below to explain the energy transfers involved in a solar panel.

In a solar panel, _____ energy is converted into _____ energy.

Some energy is converted into _____ energy and lost to the surroundings.   **[3]**

2 A student tested four different types of fleece, **J**, **K**, **L** and **M**, to find out which would make the warmest jacket.
Each type of fleece was wrapped around a can.
The can was then filled with hot water.
The temperature of the water was recorded every 2 minutes for a 20-minute period.

The graph in **Figure 2** shows the student's results.

Figure 1

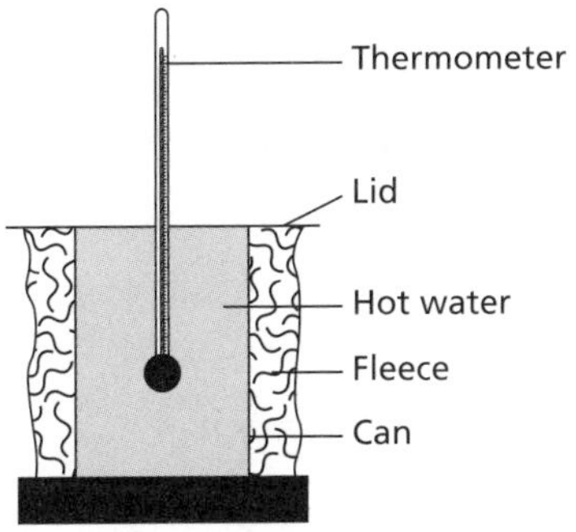

Thermometer

Lid

Hot water

Fleece

Can

Figure 2

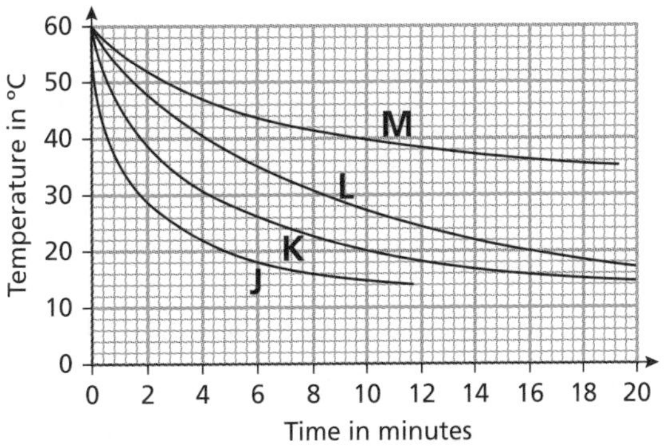

a) To be able to compare the results, it was important to use the same volume of water in each test.

Give **two** other variables that should have been kept the same in each test.

_____   **[2]**

b) Which type of fleece, **J**, **K**, **L** or **M**, should the student recommend for making a ski jacket?
You must explain your answer by making reference to the thermal conductivity of the different fleeces.

_____

_____

_____   **[3]**

Total Marks _____ / 8

Waves and Wave Properties

1 A wave machine in a swimming pool generates waves with a frequency of 0.5Hz.

a) What does a frequency of 0.5Hz mean?

.... **[1]**

b) Give the equation that links the frequency, speed and wavelength of a wave.

.... **[1]**

c) The swimming pool is 50m long.
It takes each wave 10 seconds to travel the length of the pool.

Calculate the wave speed.

Answer: .... **[2]**

d) Use your answers to parts **b)** and **c)** to calculate the wavelength of the waves.

Answer: .... **[2]**

e) One section of the swimming pool is designed for young children and has much shallower water.
A parent notices that the waves get closer together when they enter this section.

What effect will this have on the wave speed?

.... **[1]**

2 Waves may be longitudinal or transverse.

Describe the differences between longitudinal and transverse waves.

....

....

....

....

....

.... **[3]**

Total Marks .... / 10

Electromagnetic Waves

1 **Figure 1** shows two beakers. Each beaker has a drawing pin inside.

Figure 1

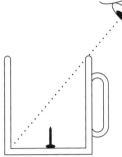

The first beaker is empty and the eye cannot see the drawing pin.
The second beaker is full of water and the drawing pin can be seen.

Explain how this is possible.
You may add a ray line to the diagram to help with your answer.

....

....

.... **[3]**

2 **Figure 2** shows the electromagnetic spectrum.

Figure 2

Radio waves	Microwaves	Infrared	Visible light	Ultraviolet	X-rays	Gamma rays

Complete the sentence below using words from the box.

amplitude   frequency   speed   wavelength

The arrow in the diagram points in the direction of increasing ....

and decreasing .... **[2]**

Total Marks .... / 5

The Electromagnetic Spectrum

1 Radio waves and visible light are electromagnetic waves that are used for communication.

a) Name another type of electromagnetic wave that is used for communication.

Answer: _____ **[1]**

b) Name an electromagnetic wave that is **not** used for communication and give one of its uses.

_____

_____ **[2]**

2 After a person is injured, a doctor will sometimes request a photograph of the patient's bones.

a) Which type of electromagnetic radiation would be used to produce the photograph?

Answer: _____ **[1]**

b) What properties of this radiation enable it to be used to photograph bones?

_____

_____

_____ **[2]**

c) Another type of electromagnetic wave is ultraviolet light, which is used in sunbeds.

i) State **one** hazard of sunbed use.

_____ **[1]**

ii) Explain why ultraviolet light is more hazardous than visible light.

_____

_____ **[2]**

iii) Despite the risks many people still regularly use sunbeds.

Suggest **two** reasons why.

_____

_____ **[2]**

Total Marks _____ / 11

An Introduction to Electricity

1 Circle the correct words to complete the sentences.

Electric **current** / **charge** is the flow of electrical **charge** / **potential**.

The **greater** / **smaller** the flow, the higher the **current** / **voltage**. **[4]**

2 The element in a set of hair straighteners has a 5A current running through it and a 230V potential difference across it.

a) Write down the equation that links potential difference, current and resistance.

Answer: .... **[1]**

b) Calculate the resistance of the element.

Answer: .... **[2]**

c) Write down the equation that links charge, current and time.

Answer: .... **[1]**

d) The straighteners are used for 2 minutes.

Calculate the charge that flows in this time.

Answer: .... **[2]**

3 Circle the correct words to complete the sentences.

Potential difference determines the amount of **energy** / **power** transferred by the charge as it passes through a component.

The **greater** / **smaller** the potential difference, the higher the **current** / **voltage** that will flow. **[3]**

4 Write the name of the component that each circuit symbol represents.

a) ⎯o⁄o⎯ Answer: .... **[1]**

b) ⊣|⋯|⊢ Answer: .... **[1]**

c) ⎯▭⎯ Answer: .... **[1]**

d) Answer: .... **[1]**

Total Marks .... / 17

Circuits and Resistance

1   The circuit in **Figure 1** is used to measure the current and potential difference of various components.

Figure 1

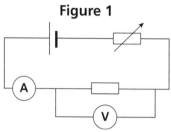

a) What is the purpose of the variable resistor?

....

.... **[2]**

b) Is the ammeter in **Figure 1** connected in series or parallel with the variable resistor?

Answer: .... **[1]**

c) Is the voltmeter in **Figure 1** connected in series or parallel with the resistor?

Answer: .... **[1]**

2   Draw **one** line from each component to the correct description.

Light dependent resistor (LDR)	Resistance decreases as temperature increases.
Thermistor	Resistance increases as the temperature increases.
Diode	Resistance decreases as light intensity increases.
Filament light	Has a very high resistance in one direction.

[3]

Total Marks .... / 7

Circuits and Power

1 This question is about a hairdryer that heats air and blows it out the front through a nozzle.

a) The hairdryer has an input power of 1600W.

If a person takes two minutes to dry their hair, how much energy has been transferred?

Answer: .... **[3]**

b) The hairdryer is 90% efficient. The remaining energy is output as sound.

How much sound energy is produced in two minutes of use?

Answer: .... **[2]**

2 **Figure 1** shows a series circuit with two resistors, **X** and **Y**.

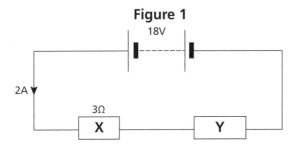

Figure 1

a) Calculate the potential difference across resistor **X**.

Answer: .... **[2]**

b) Use your answer to part **a)** to work out the potential difference across component **Y**.

Answer: .... **[2]**

c) Calculate the total resistance of the circuit.

Answer: .... **[2]**

Total Marks .... / 11

Domestic Uses of Electricity

1 A battery is connected to an oscilloscope and the trace in **Figure 1** is produced.

Figure 1

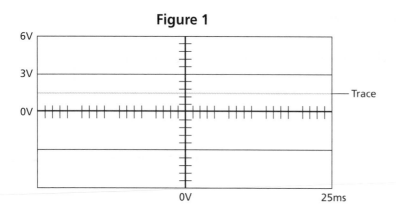

a) Use the trace to determine the potential difference of the battery and the type of current.

.... **[2]**

b) The battery is replaced by the mains supply and the trace in **Figure 2** is recorded by the oscilloscope.

Figure 2

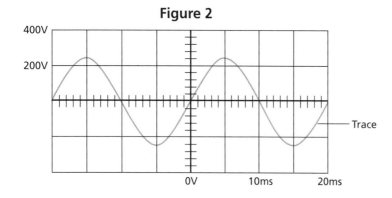

Use the trace to determine the potential difference and the type of current.

.... **[2]**

2 An appliance is switched off but the power cable is connected to the mains.

Explain how a live wire can still be dangerous.

....

....

....

....

.... **[4]**

Total Marks .... / 8

Electrical Energy in Devices

1 An electric blender transfers electrical energy into kinetic, heat and sound energy.

a) What is the useful energy output?   Answer: _____ [1]

b) What happens to the waste energy produced?

_____

_____ [2]

2 Table 1 gives some information about an electric drill.

Table 1

Energy Input	
Useful Energy Output	
Wasted Energy	Heat and Sound
Power Rating	500W

a) Complete **Table 1** by adding the missing types of energy.   [2]

b) Which of the following statements about the energy from the drill is **incorrect**?
Tick **one** box.

It spreads out and becomes more difficult to use.   ☐

It disappears.   ☐

It makes the surroundings warmer.   ☐   [1]

c) How much energy does the drill use per second?   Answer: _____ [1]

3 The National Grid distributes electricity from power stations to consumers.
The voltage across the overhead cables of the National Grid is much higher than the output
voltage from the power station generators.

Explain how this achieved and why it is important.

_____

_____

_____

_____ [4]

Magnetism and Electromagnetism

1 The full name for the north pole of a magnet is the 'north-seeking pole'.

Explain what is meant by this.

....

.... **[2]**

2 The north pole of a permanent magnet is moved close to the north pole of another permanent magnet.

a) What would you expect to happen?

.... **[1]**

b) A piece of iron is moved close to the north pole of a permanent magnet.

What would you expect to happen?

.... **[1]**

3 **Figure 1** shows an electric bell.

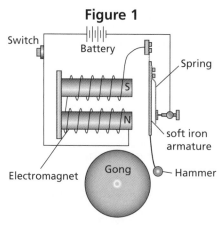

Figure 1

Describe what happens when the switch is pressed.

....

....

....

.... **[5]**

Total Marks .... / 9

The Motor Effect

1   A current-carrying wire passes through a magnetic field at right-angles to the field and experiences a force.

The length of wire in the field is 5cm, the current is 2A and the magnetic field is 0.3mT.

a) Calculate the force on the wire.
Use the correct equation from the Physics Equation Sheet on page 169.

Answer: _____ **[2]**

b) The magnets are rearranged so that the current flowing in the wire is parallel to the field lines.

How will this affect the force on the wire?

_____ **[1]**

2   **a)** The left hand rule can be used to identify the direction of the force acting on a conductor carrying a current in a magnetic field.

State what is indicated by each of the fingers in the left hand rule.

_____

_____

_____ **[3]**

b) State **two** ways of increasing the speed of rotation of an electric motor.

_____

_____ **[2]**

c) State **two** ways of reversing the direction of an electric motor.

_____

_____ **[2]**

Total Marks _____ / 10

Particle Model of Matter

1 Heating a substance can cause it to change state from a solid to a liquid or from a liquid to a gas.

a) What is meant by 'specific latent heat of fusion'?

....

.... **[2]**

b) While a kettle boils, 0.012kg of water changes to steam.

Calculate the amount of energy required for this change.
Use the correct equation from the Physics Equation Sheet on page 169.
Specific latent heat of vaporisation of water = 2.3 × 10⁶J/kg

Answer: .... **[2]**

2 The graph in **Figure 1** shows how temperature varies with time as a substance cools. The graph is **not** drawn to scale.

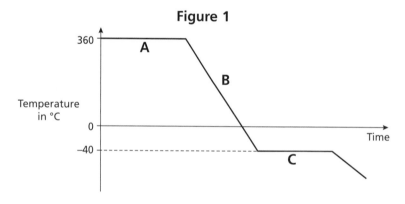

Figure 1

a) Explain what is happening to the substance in section **A** of the graph.

....

.... **[2]**

b) Explain what is happening to the substance in section **B** of the graph.

....

....

.... **[2]**

Total Marks .... / 8

Atoms and Isotopes

1 Atoms contain three types of particle.

a) Complete the table to show the relative charges of the subatomic particles.

Particle	Relative Charge
Electron	
Neutron	
Proton	

[3]

b) A neutral atom has no overall charge.

Explain why in terms of its particles.

....

....

....

[2]

c) Complete the sentences below.

An atom that loses or gains an electron becomes an ....

If it loses an electron, it has an overall .... charge. [2]

2 In the early part of the 20th century, Rutherford, Geiger and Marsden investigated the paths taken by positively charged alpha particles into and out of a very thin piece of gold foil. **Figure 1** shows the paths of three alpha particles.

Explain the different paths, **A**, **B** and **C**, of the alpha particles.

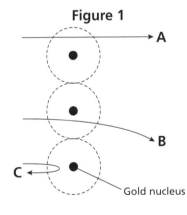

Figure 1

....

....

....

[3]

Total Marks .... / 10

Nuclear Radiation

1   Here is some information about potassium.

   Potassium is a metallic element in Group 1 of the Periodic Table.
   It has an atomic number of 19.
   Its most common isotope is potassium-39, $^{39}_{19}K$.
   Another isotope, potassium-40, $^{40}_{19}K$, is a radioactive isotope.

   a) What is meant by 'radioactive isotope'?

   .... [1]

   b) During radioactive decay, atoms of potassium-40 change into atoms of calcium-40.
      Calcium-40 has an atomic number of 20 and a mass number of 40.

      What type of radioactive decay has taken place?

      Answer: .... [1]

   c) Potassium-39 does not undergo radioactive decay.

      What does this tell us about potassium-39?

      .... [1]

   d) Sodium-24 is another radioactive isotope.
      It decays by gamma emission.

      Give the name of the element formed when this decay takes place.

      Answer: .... [1]

2   Give the unit that is used to measure the activity of a radioactive isotope.

      Answer: .... [1]

3   List the decay mechanisms, **alpha**, **beta** and **gamma**, in order of penetrating power.
   Start with the most penetrative.

   .... [1]

   Total Marks .... / 6

Half-Life

1 Iodine is found naturally in the world and is essential to life.
It is used by the thyroid gland for the production of essential hormones.
Iodine-127 is not radioactive but iodine-131 is.
Iodine-131 has as a half-life of 8 days.

a) During the Chernobyl nuclear disaster in 1986, an explosion caused a large quantity of the isotope iodine-131 to be released into the atmosphere.

Is iodine-131 from the disaster still a threat to us today?
Explain your answer.

[3]

b) A sample of iodine-131 has a count-rate of 256 counts per minute.

Work out the count-rate of the sample after 24 days.

Answer: _____ [2]

c) The Isotope, Caesium-137, was also released during the disaster.
Ceasium-137 has a half-life of 30 years.

In what year will the activity of the caesium released in the disaster have fallen to $\frac{1}{8}$ of the initial activity?

Answer: _____ [2]

d) Potassium has an atomic number of 19. It decays into calcium-40 by emitting an electron from the nucleus.

Produce a balanced decay equation for the decay of potassium-40 into caesium-40.

[3]

Total Marks _____ / 10

Notes

Collins

GCSE
COMBINED SCIENCE

H

Biology: Paper 1 Higher Tier

Materials

Time allowed: 1 hour 15 minutes

For this paper you must have:

- a ruler
- a calculator.

Instructions

- Answer **all** questions in the spaces provided.
- Do all rough work on the page. Cross through any work you do not want to be marked.

Information

- There are **70** marks available on this paper.
- The marks for each question are shown in brackets [ ].
- You are expected to use a calculator where appropriate.
- You are reminded of the need for good English and clear presentation in your answers.
- When answering questions 03.2 and 05.3 you need to make sure that your answer:
  – is clear, logical, sensibly structured
  – fully meets the requirements of the question
  – shows that each separate point or step supports the overall answer.

Advice

- In all calculations, show clearly how you work out your answer.

01 Gonorrhoea is a disease caused by a microorganism.

Figure 1 shows the microorganism that causes gonorrhoea.

Figure 1

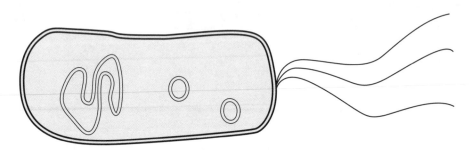

X   Y

01.1 What type of microorganism is this?

Tick **one** box.

Bacterium   ☐

Fungus   ☐

Protist   ☐

Virus   ☐   **[1 mark]**

01.2 The magnification of **Figure 1** is ×14 000.

The length of the microorganism is shown by line **XY**.

$$\text{magnification} = \frac{\text{size of image}}{\text{size of real object}}$$

What is the real length of the microorganism in millimetres?

Real length = _____ mm   **[2 marks]**

01.3 The microorganism can reproduce once every 40 minutes in ideal conditions.

Calculate how many microorganisms could be produced from one microorganism in 4 hours.

Answer = _____ microorganisms **[2 marks]**

Turn over for the next question

02   **Figure 2** shows the number of people who tested positive for salmonella food poisoning each month in one year.

Figure 2

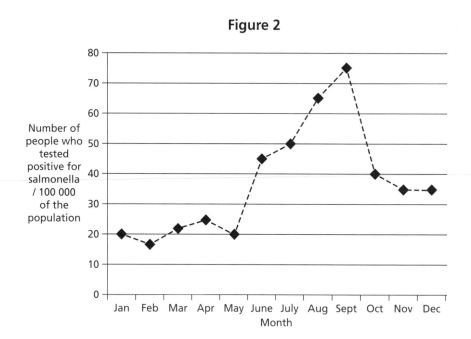

02.1 Describe the symptoms that the people who tested positive for salmonella were likely to show.

....

....   **[2 marks]**

02.2 Using your knowledge of how salmonella is passed on, suggest reasons for the pattern shown in **Figure 2**.

....

....

....   **[2 marks]**

03   Doctors are hoping that in the future they will be able to treat heart disease by injecting stem cells into a damaged heart.

Figure 3 shows where the stem cells are injected.

Figure 3

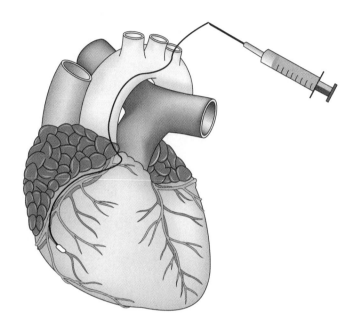

03.1 Which chamber of the heart are the cells being injected into?

Tick **one** box.

Left ventricle

Right ventricle

Left atrium

Right atrium   **[1 mark]**

Question 3 continues on the next page

03.2 Describe the cause of heart disease.

Include some of the risk factors associated with the disease.

[6 marks]

The stem cells for this treatment can be taken from an embryo that has been produced by cloning the patient's own cells.

03.3 Explain the benefit of using stem cells from an embryo cloned from the patient's own cells.

[2 marks]

03.4 Suggest why some people may object to this treatment.

[2 marks]

04   Penicillin is an antibiotic drug.

04.1 Draw **one** line to link the person who discovered penicillin to the organism that it comes from.

person	organism
Mendel	fungus
Fleming	willow
Wallace	foxglove

[2 marks]

04.2 Some strains of bacteria have developed resistance to penicillin.

As more bacteria develop resistance, there is more pressure on scientists to produce new drugs.

Here are four steps carried out in the testing of new drugs.

Double-blind trials on patients	
Varying doses given to healthy volunteers	
Testing on live animals	
Low doses given to healthy volunteers	

Write the numbers **1**, **2**, **3** or **4** in the boxes to show the order of these steps in drug testing.

[3 marks]

Turn over for the next question

05 Sunflower plants, such as the ones shown in **Figure 4**, can grow well in the UK.

They often grow up to three metres tall.

Figure 4

05.1 Suggest why it is an advantage for sunflowers to be taller than the other plants growing around them.

....

....

....

.... **[3 marks]**

05.2 Growing tall means that sunflower plants have to transport water several metres up to the leaves from the roots.

Describe how they do this.

....

....

....

.... **[3 marks]**

Figure 5 shows an aloe plant.

Aloe is a plant that grows in desert areas.

In these areas there is little food for animals and very little water.

Figure 5

Table 1 gives some differences between the leaves of aloe plants and sunflowers.

Table 1

Plant	Thickness of Waxy Cuticle (micrometres)	Number of Stomata (per mm^2)
aloe	14.8	25
sunflower	6.1	150

05.3 Explain how aloe is adapted to living in desert conditions.

Use **Figure 5** and the data in **Table 1**.

[6 marks]

Turn over for the next question

06 Racehorses and athletes are both trained to run in races.

Figure 6

Horses can be tested to see how fit they are.

A horse's heart rate is measured when it is running at different speeds.

Some results for a horse are shown in **Table 2**.

Table 2

Speed of the Horse (Kilometres per Hour)	Heart Rate (Beats per Minute)
30	162
35	170
39	180
45	192

06.1 Plot the data from **Table 2** on the graph in **Figure 7**.

Finish the graph by drawing the line of best fit.

Figure 7

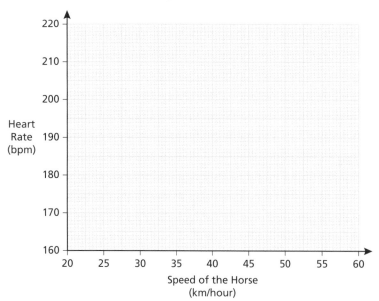

[4 marks]

06.2 Describe the pattern shown in the graph.

....

....

.... **[2 marks]**

06.3 Above 200 heart beats per minute, a horse starts to use **anaerobic** respiration.

Use the graph to estimate the maximum speed at which this horse can run **without** using anaerobic respiration.

Show on the graph how you work out your answer.

Maximum speed = ... km/hour **[2 marks]**

06.4 Write the word equation to show the reaction for **anaerobic** respiration in horses and athletes.

.... **[2 marks]**

06.5 Explain why athletes and horses run better when they use **aerobic** rather than **anaerobic** respiration.

....

....

.... **[3 marks]**

Turn over for the next question

07 A student investigates the digestion of fats (lipids) by the enzyme lipase.

They find that when lipase digests fats, the pH of the solution changes from pH 8 to pH 6.

07.1 Why did the digestion of the fats change the pH of the solution?

Tick **one** box.

Amino acids are formed ☐

Fatty acids are formed ☐

Fats are alkaline ☐

Glucose is formed ☐   **[1 mark]**

The student sets up three test tubes containing various liquids.
Figure 8 shows the contents of the tubes.

Figure 8

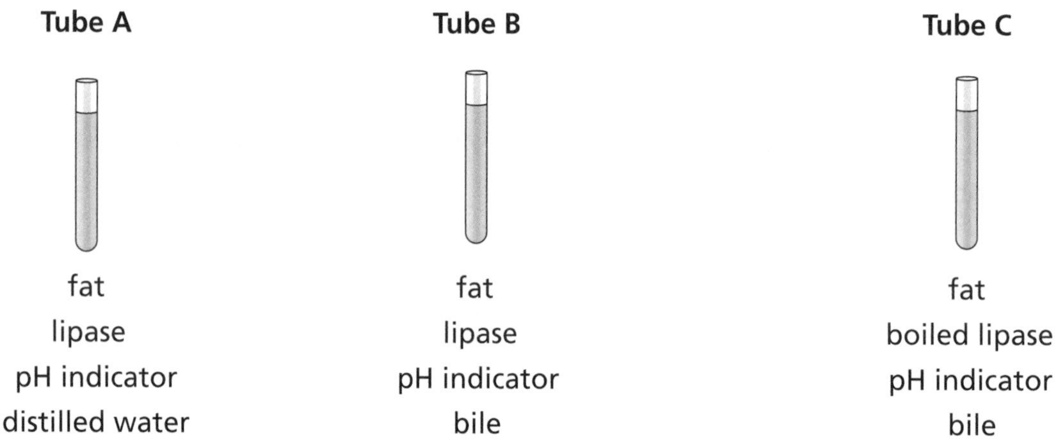

Tube A	Tube B	Tube C
fat	fat	fat
lipase	lipase	boiled lipase
pH indicator	pH indicator	pH indicator
distilled water	bile	bile

The student times how long it takes for the indicator to change colour after the lipase is added.

The results are shown in **Table 3**.

Table 3

Tube	Time Taken (minutes)
A	8
B	1
C	No change after 20 minutes

07.2 The indicator in tube **B** changes colour much faster than the indicator in tube **A**.

Explain why.

[3 marks]

07.3 Explain why the colour did not change in tube **C**.

[2 marks]

07.4 Explain why lipase can digest fats (lipids) but amylase or protease cannot.

[3 marks]

Turn over for the next question

08 A student wants to find out the concentration of the cell contents of potato.
They design an experiment involving osmosis.

08.1 What is meant by the term 'osmosis'?

....

....

.... **[3 marks]**

The student cuts cylinders from a potato and weighs each cylinder.
They then place each cylinder in a test tube.
Each test tube contains a different concentration of sugar solution.
After several hours the student removes the cylinders from the solutions and reweighs them.
They then calculate the percentage change in mass for each cylinder.

Figure 9 shows the student's results.

Figure 9

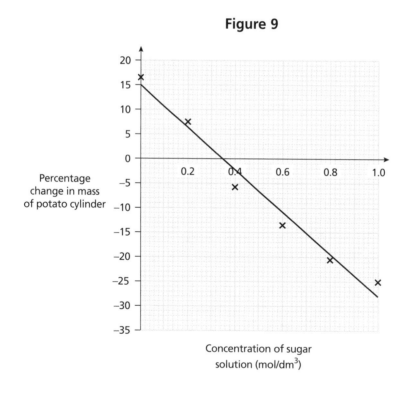

08.2 The potato cylinders in 0.4 to 1.0 mol/dm³ sugar solution all lost mass.

Explain why.

_____

_____

_____ **[3 marks]**

08.3 Use **Figure 9** to estimate the concentration of the cell contents of the student's potato.

Concentration = _____ mol/dm³ **[1 mark]**

08.4 **Table 4** shows possible variables in the student's experiment.

Put the letter **I**, **D** or **C** in the table to show the type of variable.

I = independent variable
D = dependent variable
C = any variables that should have been controlled

Table 4

The concentration of the sugar solution	
The volume of the sugar solution	
The change in mass of the potato cylinder	
The time that each cylinder was left to soak	

[4 marks]

END OF QUESTIONS

There are no questions printed on this page

Collins

GCSE
COMBINED SCIENCE
Biology: Paper 2 Higher Tier

H

Materials

Time allowed: 1 hour 15 minutes

For this paper you must have:

- a ruler
- a calculator.

Instructions

- Answer **all** questions in the spaces provided.
- Do all rough work on the page. Cross through any work you do not want to be marked.

Information

- There are **70** marks available on this paper.
- The marks for each question are shown in brackets [ ].
- You are expected to use a calculator where appropriate.
- You are reminded of the need for good English and clear presentation in your answers.
- When answering questions 02.4 and 08.3 you need to make sure that your answer:
  - is clear, logical, sensibly structured
  - fully meets the requirements of the question
  - shows that each separate point or step supports the overall answer.

Advice

- In all calculations, show clearly how you work out your answer.

01   This question is about skin sensitivity.

Tim presses Kate's skin with two pin heads, as shown in **Figure 1**.

Sometimes Tim presses both pin heads on to Kate's skin and sometimes just one.
Every time he presses her skin, Kate says the number of points she feels.
Tim writes down how many times she calls the correct number of points.
He does this 20 times.

Tim does this experiment with the pins different distances apart and on three parts of the body.
The graph in **Figure 2** shows his results.

Figure 1

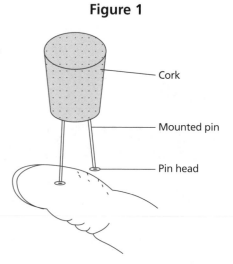

Cork

Mounted pin

Pin head

Figure 2

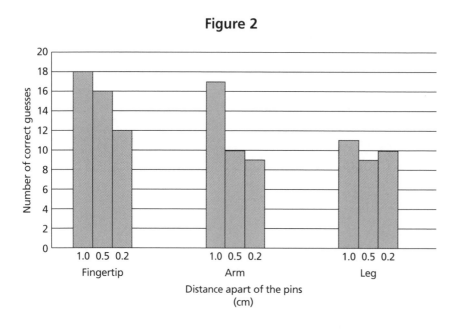

01.1 Name the part of Kate's body where her skin was most sensitive to the stimulus.

_____ **[1 mark]**

01.2 Explain why Tim touched Kate's skin 20 times in each trial.

_____

_____ **[2 marks]**

01.3 What do the results from Kate's arm show?

_____

_____ **[2 marks]**

01.4 Kate's response to the pins was not a reflex action.

Explain how you can tell this.

_____

_____ **[2 marks]**

01.5 **Figure 3** shows a reflex arc.

Figure 3

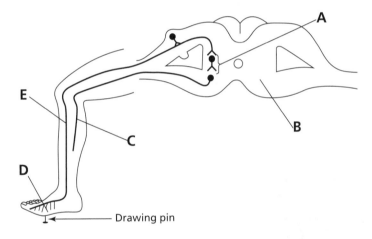

Complete **Table 1** by matching a letters on the diagram to each part of the reflex arc.
The first one has been done for you.

Table 1

Part of Reflex Arc	Letter
relay neurone	A
receptor	
sensory neurone	
spinal cord	
motor neurone	

[3 marks]

Turn over for the next question

02   Polar bears (scientific name *Ursus maritimus*) live in the Arctic.
They need to move about on the ice and they only eat seals.

Figure 4

02.1 Describe how polar bears are adapted for hunting and eating seals on ice.

....

.... **[2 marks]**

Alaskan bears (*Ursus arctos)* live in Alaska, south of the Arctic.
They catch fish for food.

Figure 5

02.2 What do their scientific names tell you about how closely related the Alaskan bear
and the polar bear are?

....

.... **[2 marks]**

02.3 The temperature in Alaska is getting warmer.

Alaskan bears are now living in some of the same areas as polar bears.

Which of these statements are true?

Tick **two** boxes.

Polar bears and Alaskan bears will compete for food. ☐

Polar bears will mate with Alaskan bears and the offspring will be a new species. ☐

Polar bears and Alaskan bears may mate and produce sterile hybrids. ☐

Polar bears and Alaskan bears cannot mate because they are in different kingdoms. ☐

The habitats of the polar bears and the Alaskan bears may overlap. ☐ **[2 marks]**

02.4 The level of carbon dioxide in the air is increasing.

Explain why scientists think this is happening and how it might be causing the temperature in the Arctic to increase.

....

....

....

....

....

....

....

.... **[6 marks]**

Turn over for the next question

03   Copper is an important element in many organisms. However, in high concentrations it is poisonous.

Figure 6 shows waste soil from a copper mine piled up next to a river.

Figure 6

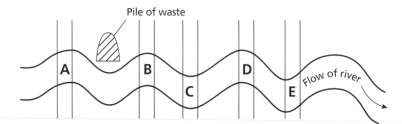

A scientist investigated copper concentrations in the river.
They took samples of water from locations **A**, **B**, **C**, **D** and **E**.
They then measured the concentration of copper in each sample.

The results are shown in **Table 2**.

Table 2

Sample Location	Concentration of Copper (Micrograms per Litre)
A	1
B	16
C	10
D	7
E	3

03.1 Suggest an explanation for the results.

....

.... **[2 marks]**

03.2 The scientist noticed that there were fewer plants growing on the river bank at location **B** than at location **A**.

Suggest an explanation for this difference.

....

.... **[2 marks]**

The scientist takes sample plants from each location and continues to grow them in the laboratory.

The scientist waters the plants regularly with water containing copper.

The survival rate of the plants is shown in **Table 3**.

Table 3

Location That Plants Were Taken From	Survival (%)
A	1
B	100
C	90
D	55
E	20

03.3 Why is there such a large difference in the survival rate of plants from location **A** compared with those from location **B**?

....

.... **[2 marks]**

03.4 The scientist explained the results by using natural selection.

Explain how natural selection could be used to explain why 100% of the plants from location **B** could survive the watering.

....

....

....

....

.... **[4 marks]**

Turn over for the next question

04 In 1866, Gregor Mendel published a paper on genetics.

Figure 7

Mendel carried out breeding experiments with pea plants.
He crossed tall pea plants with dwarf pea plants.
He collected the seeds and planted them.
He then self-pollinated the plants that grew and again planted the seeds that were made.

His results are shown in **Figure 8**.

Figure 8

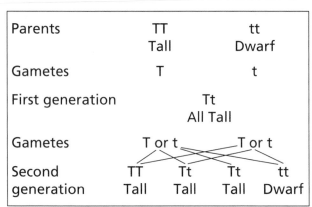

Parents	TT Tall	tt Dwarf
Gametes	T	t
First generation	Tt All Tall	
Gametes	T or t	T or t
Second generation	TT Tt Tall Tall	Tt tt Tall Dwarf

04.1 Complete the sentences about Mendel's results.

Use words from the box.

dominant   dwarf   genotypes   heterozygous   homozygous   phenotypes   recessive   tall

The genotype of all the plants in the first generation is _____ .

The phenotype of these plants is _____ .

This is because the tall allele is _____ over the dwarf allele.

The second generation contains plants that have three

different _____ .

[4 marks]

04.2 To test his ideas, Mendel crossed a plant from the first generation with a dwarf plant.

Draw a genetic diagram to show this cross.
Write down the ratio of phenotypes that this cross would produce.

Ratio: _____ **[4 marks]**

Mendel then self-pollinated the plants that were Tt **or** TT.
He looked at the first 10 offspring.
If they were **all** tall, he said that the parent plant was TT.
If **any** were dwarf, he said that the parent plant was Tt.

04.3 Explain why Mendel's assumptions may have been wrong in a small number of cases.

_____

_____ **[2 marks]**

Turn over for the next question

05 Here is an advert of a DNA testing kit.

> **BABY TESTING KIT**
>
> Do you want to know the sex of your baby before it is born?
> This testing kit can answer that question.
> When a woman is pregnant, a very small amount of the baby's DNA is in the
> mother's blood.
> This means a small drop of the mother's blood can be tested to see if it has any DNA
> from a Y chromosome.
> This will tell you if you are going to have a boy or a girl.

05.1 The test involves finding out if any DNA in the mother's blood is from a Y chromosome.

Explain how this can be used to tell the sex of the baby.

....

.... **[2 marks]**

05.2 At present, scientists do this test on cells taken directly from the baby inside the
mother's uterus.
They hope that in the future they will be able to carry out the same tests using
samples of the mother's blood.

Suggest why scientists think that this would be a better method.

....

.... **[2 marks]**

06 *Acetabularia* is a single-celled organism that lives in the sea.

It has an unusual shape, as shown in **Figure 9**.

Figure 9

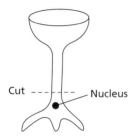

Cut ----------- Nucleus

06.1 When *Acetabularia* reproduces, the nucleus and then the cytoplasm divides into two.

What is this type of reproduction called?

_____ **[1 mark]**

06.2 *Acetabularia* produced by this type of reproduction always have the same shaped cap as the parent.

Why is this?

_____ **[1 mark]**

06.3 A scientist called Hammerling performed some experiments on *Acetabularia*.
He cut the organism into two as shown on **Figure 9**.
He found that the bottom part of the organism survived and grew a new top (cap).

Explain these results.

_____

_____

_____

_____ **[3 marks]**

Turn over for the next question

07   This question is about the causes of infertility in humans.

Table 4 shows some of the problems that cause infertility.

Table 4

Problem	Percentage of Infertile Couples with this Problem	Percentage Success Rate of Treatment
blocked fallopian tubes	13	20
irregular ovulation	16	75
no ovulation	7	95
low sperm production	15	10
no sperm production	21	10
unknown cause	28	-

Use the information in **Table 4** and your biological knowledge to answer the following questions.

07.1 What is the total percentage of infertile couples where the problem is known to be in the **male**?

Percentage = ... **[1 mark]**

07.2 Treatment of which problem produces the largest **number** of pregnant women among the affected infertile couples?

Explain how you worked out your answer.

....

.... **[2 marks]**

Figure 10 shows some of the hormones released from the pituitary gland and some of the effects that they have on the ovary.

Figure 10

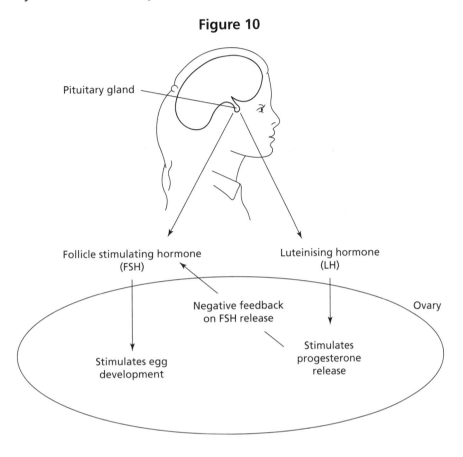

07.3 Progesterone has a negative feedback effect on FSH.

What does this mean?

_____

[1 mark]

07.4 Some women with fertility problems are injected with a drug that prevents the pituitary gland from making LH.

Explain why this might help them to get pregnant.

_____

_____

_____

[3 marks]

08   This question is about hormones and homeostasis.

08.1 What is a hormone?

....

.... **[2 marks]**

08.2 **Table 5** contains information about three different hormones.

Table 5

Hormone	Gland that Releases Hormone	Function of Hormone
		controls how much water is reabsorbed in the kidney
thyroxine		
	adrenal gland	

Complete the table by writing in the six blank boxes.   **[6 marks]**

The article in **Figure 11** appeared in a recent newspaper.

Figure 11

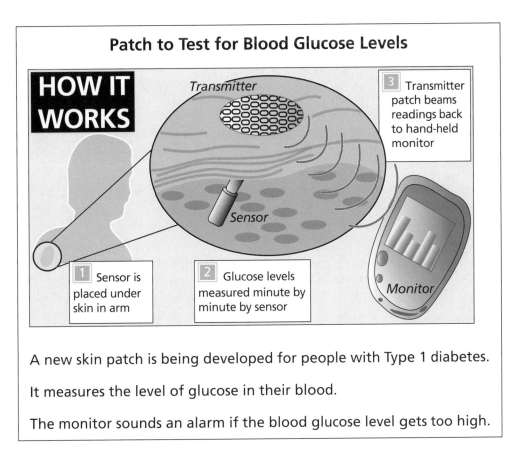

Patch to Test for Blood Glucose Levels

HOW IT WORKS

Transmitter

3 Transmitter patch beams readings back to hand-held monitor

Sensor

1 Sensor is placed under skin in arm

2 Glucose levels measured minute by minute by sensor

Monitor

A new skin patch is being developed for people with Type 1 diabetes.

It measures the level of glucose in their blood.

The monitor sounds an alarm if the blood glucose level gets too high.

08.3 It is very important for people with Type 1 diabetes to know if their blood glucose level gets too high.

Explain why.

_____

_____

_____

_____

_____

[4 marks]

END OF QUESTIONS

There are no questions printed on this page

Collins

GCSE
COMBINED SCIENCE

H

Chemistry: Paper 1 Higher Tier

Materials

Time allowed: 1 hour 15 minutes

For this paper you must have:

- a ruler
- a calculator
- the periodic table (see page 170).

Instructions

- Answer **all** questions in the spaces provided.
- Do all rough work on the page. Cross through any work you do not want to be marked.

Information

- There are **70** marks available on this paper.
- The marks for each question are shown in brackets.
- You are expected to use a calculator where appropriate.
- You are reminded of the need for good English and clear presentation in your answers.
- When answering question 05.2 you need to make sure that your answer:
  - is clear, logical, sensibly structured
  - fully meets the requirements of the question
  - shows that each separate point or step supports the overall answer.

Advice

- In all calculations, show clearly how you work out your answer.

Chemistry Practice Exam Paper 1

01 This question is about the atomic model.

Figure 1 shows a model of a helium atom.

Figure 1

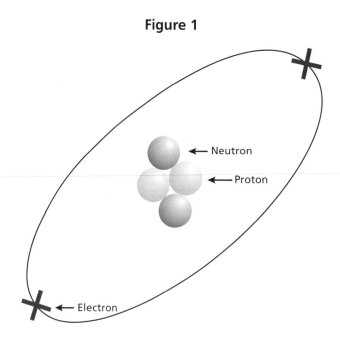

01.1 What are the relative electric charges on the particles in an atom?

Tick **one** box.

Proton	Neutron	Electron	
0	−1	+1	☐
+1	0	−1	☐
−1	0	0	☐
+1	+1	−1	☐ [1 mark]

01.2 A helium atom has an overall neutral charge.

State why.

_____

_____ [1 mark]

01.3 What is the mass number and atomic number of helium?

Tick **one** box.

Mass Number	Atomic Number		
2	2	☐	
6	2	☐	
6	4	☐	
4	2	☐	[1 mark]

01.4 Why is helium an unreactive element?

Tick **one** box.

It has an equal number of protons and neutrons. ☐

Elements with two electrons in their outer shell are unreactive. ☐

It is a gas at room temperature and pressure. ☐

It has a full outer shell of electrons. ☐ [1 mark]

01.5 An isotope of helium has only one neutron in its atoms.

Which statements about how an atom of this isotope compares to the atom in **Figure 1** are true?

Tick **two** boxes.

It has the same atomic number. ☐

It has a higher mass. ☐

It has a different atomic number. ☐

It has a different mass number. ☐

It has the same mass number. ☐ [2 marks]

Turn over for the next question

02 Iron is found in the Earth as the compound iron oxide (Fe_2O_3).

02.1 How many atoms are in one molecule of Fe_2O_3?

Answer: _____ **[1 mark]**

Iron is extracted from iron(III) oxide using carbon in the following reaction:

$2Fe_2O_3 + 3C \rightarrow 4Fe + 3CO_2$

02.2 Calculate the relative formula mass (M_r) of carbon dioxide (CO_2).

Relative atomic masses (A_r): carbon = 12; oxygen = 16

Answer: _____ **[1 mark]**

02.3 Both oxidation and reduction take place when iron is extracted from iron(III) oxide.

Draw **one** line from each type of change to the name of the substance that is changed in that way.

Change

| oxidation |

| reduction |

Substance

| iron |

| iron(III) oxide |

| carbon |

| carbon dioxide | **[2 marks]**

02.4 Gold does not have to undergo an extraction process.

Explain why.

....

....

....

....　**[2 marks]**

Turn over for the next question

Chemistry Practice Exam Paper 1

03   **Table 1** shows some of the properties of elements in Group 7 of the periodic table.

Table 1

Element	Density (g/cm³)	Melting point (°C)	Boiling point (°C)
Fluorine	0.0017	−219.6	−188.1
Chlorine	0.0032	−101.5	−34.0
Bromine	3.1028	−7.3	58.8
Iodine	4.9330	113.7	184.3

03.1 State **one** trend in the properties of the Group 7 elements as shown in **Table 1**.

....

.... **[1 mark]**

03.2 Astatine is found in Group 7, below iodine.

Use the information in **Table 1** to estimate the melting point of astatine.

Answer: .... °C   **[1 mark]**

Group 7 elements all exist as molecules containing two atoms.

03.3 Complete the dot and cross diagram in **Figure 2** to show the covalent bonding in a molecule of chlorine.

Show the outer shell electrons only.

Figure 2

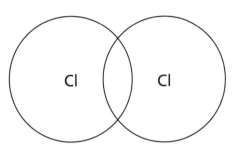

[2 marks]

03.4 What is the formula for a molecule of bromine?

Answer: .... [1 mark]

03.5 Which statement correctly describes what happens when a Group 7 element boils?

Tick **one** box.

Covalent bonds form ☐

Intermolecular forces form ☐

Covalent bonds break ☐

Intermolecular forces break ☐ [1 mark]

Question 3 continues on the next page

03.6 The Group 7 elements all react with Group 1 elements to form ionic compounds.

Explain why all the Group 7 elements share this chemical property.

....

.... **[1 mark]**

03.7 The Group 1 element sodium reacts with chlorine to form sodium chloride.
Figure 3 shows the outer electrons in an atom of sodium and in an atom of chlorine.

Figure 3

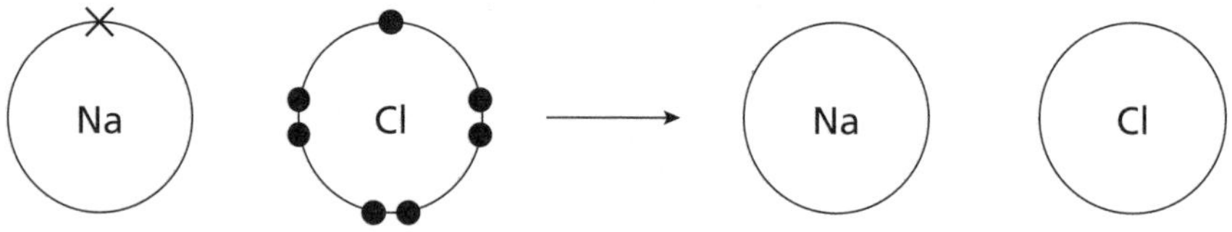

Complete the diagram to show the arrangement of electrons in sodium chloride.
You should give the formula of each ion formed. **[5 marks]**

03.8 The Group 7 elements become less reactive as you go down the group.

Explain why.

....

....

....

....

.... **[4 marks]**

04 A student was asked to produce copper(II) sulfate crystals.
They used the method shown in **Figure 4.**

Figure 4

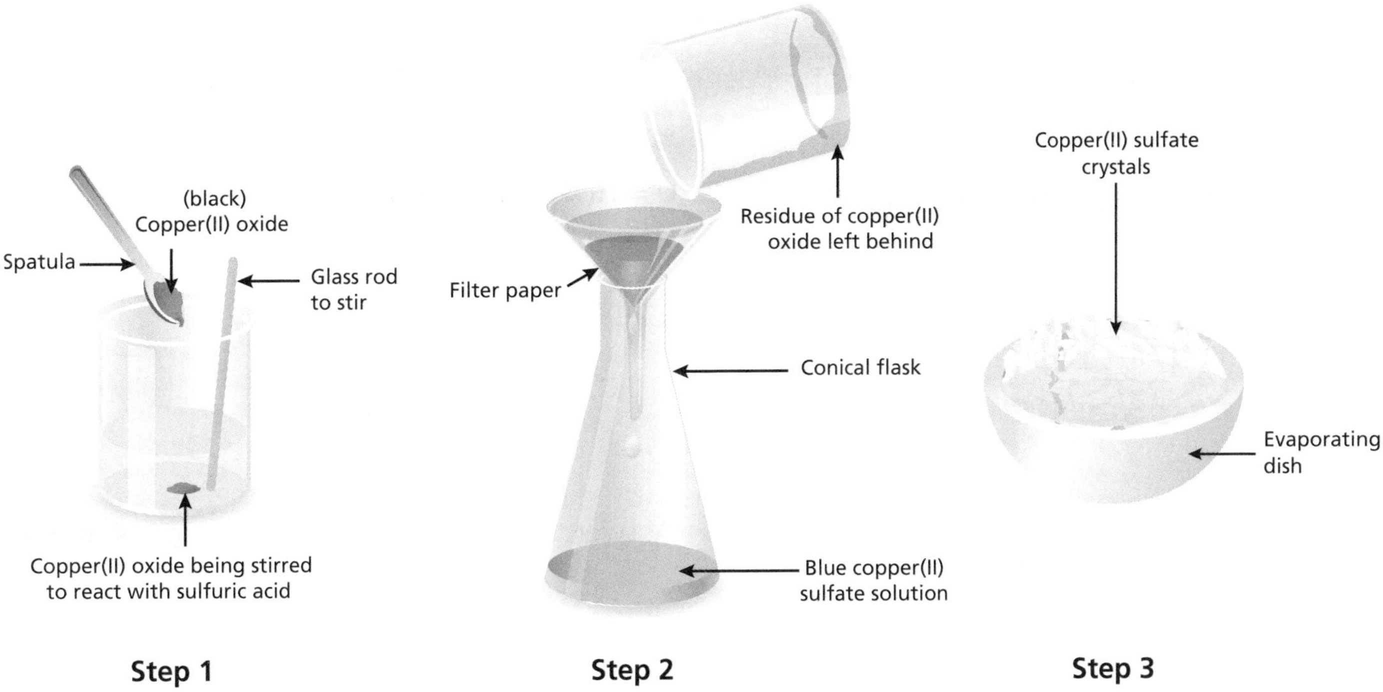

Step 1   Step 2   Step 3

04.1 Give the name of the separation process used in **Step 2** and explain
why it was used.

[2 marks]

04.2 Identify one hazard in **Step 1** or **Step 2** and suggest a method of
reducing the risk.

Hazard:

Way of reducing the risk:

[2 marks]

Question 4 continues on the next page

The equation for the reaction is:

$CuO(s) + H_2SO_4(aq) \rightarrow CuSO_4.5H_2O \, (aq) + H_2O(l)$

04.3 The student used 5.2 g of copper(II) oxide.

Calculate the maximum mass of copper(II) sulfate crystals ($CuSO_4.5H_2O$) that can be produced.

Relative atomic masses (A_r): H = 1; C = 12; O = 16; S = 32; Cu = 63.5

Answer: .... **g**   **[4 marks]**

05 **Figure 5** shows the structure of graphite.

Figure 5

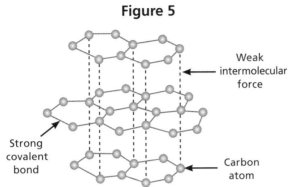

Weak intermolecular force

Strong covalent bond

Carbon atom

05.1 Graphite and diamond are both giant covalent compounds.

State **one difference** and **one similarity** between the structures of diamond and graphite.

Similarity: ....

Difference: .... **[2 marks]**

05.2 Graphene is a single layer of graphite.
Its properties include being an electrical conductor, strong and transparent.
In the future, it could be used to make touch-screens for electronic devices like mobile phones.

Explain why graphene has these properties in terms of its structure and why it is a good choice of material for a touch-screen.

....

....

....

....

....

....

....

.... **[6 marks]**

Turn over for the next question

06 A student was asked to carry out the electrolysis of copper(II) sulfate solution using inert graphite electrodes.

06.1 Complete **Figure 6** to show the apparatus they should use.
Label the following on your completed diagram:

- Anode
- Cathode.

Figure 6

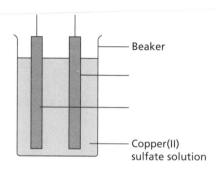

Beaker

Copper(II)
sulfate solution

[3 marks]

06.2 Hydrogen (H^+) ions and copper (Cu^{2+}) ions are both attracted to the cathode but only one product is formed.

Predict the name of the product formed at the cathode.
Explain why only this product is formed.

....

.... **[2 marks]**

06.3 Complete the balanced half equation for the reaction at the anode.

$4OH^-(aq) \rightarrow$ .... $H_2O(l) + O_2(g) +$ .... e^- **[2 marks]**

06.4 Is the reaction at the anode an example of reduction or oxidation?
Give a reason for your answer.

....

....

.... **[2 marks]**

07 A student investigated how the mass of magnesium changes when it reacts with dilute hydrochloric acid.

Figure 7 shows the apparatus they used.

Figure 7

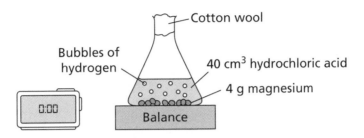

07.1 Complete the balanced symbol equation for the reaction by adding state symbols.

Mg........... + 2HCl........... → MgCl$_2$(aq) + H$_2$........... [1 mark]

07.2 State **one** function of the cotton wool.

....

.... [1 mark]

Table 2 shows the student's results.

Table 2

Time in s	Total loss of mass in g
0	0.00
10	1.25
20	2.30
30	3.00
40	3.45
50	3.70
60	3.87
70	4.00
80	4.00
90	4.00

Question 7 continues on the next page

07.3 On **Figure 8**:

- Plot the results from **Table 2**.
- Draw a line of best fit.

Figure 8

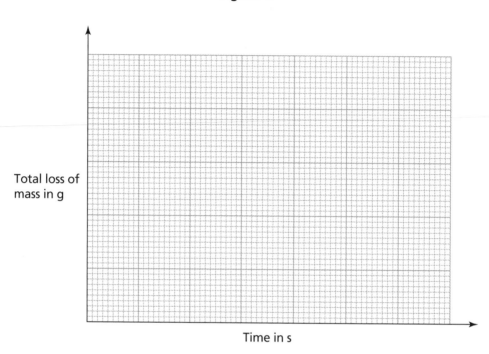

Total loss of mass in g

Time in s

[4 marks]

07.4 At what time did the mass loss stop changing?

Answer: .... s   [1 mark]

07.5 The student observed that there was still magnesium in the flask at this time.

Explain why the mass loss stopped changing.

....

.... [1 mark]

07.6 Explain, in terms of particles, why the mass changes during the reaction.

....

....

....

.... [2 marks]

07.7 Zinc is less reactive than magnesium.

Sketch a line on the graph in **Figure 8** to show the results you would expect if the experiment was repeated using zinc of the same surface area.

Label this line **A**.   [2 marks]

Turn over for the next question

08   This question is about the reaction of hydrogen with oxygen.
The equation for the reaction is:

$2H_2(g) + O_2(g) \rightarrow 2H_2O(g)$

Figure 9 shows the displayed formulae for the reaction.

Figure 9

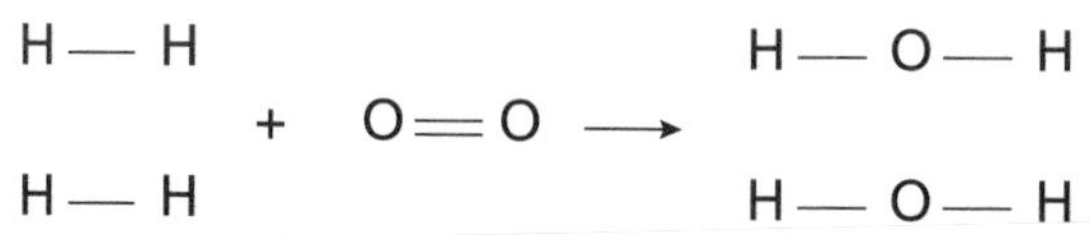

```
H — H                          H — O — H
          +    O = O  ⟶
H — H                          H — O — H
```

The bond enthalpies for the reaction are shown in Table 3.

Table 3

	H–H	O=O	O–H
Energy (kJ/mol)	432	495	467

08.1 Calculate the energy transferred in the reaction.

Answer: _____ kJ/mol **[3 marks]**

08.2 Explain, in terms of bond energies, why the reaction is exothermic.

_____

_____

_____

[2 marks]

END OF QUESTIONS

Collins

GCSE
COMBINED SCIENCE
Chemistry: Paper 2 Higher tier

H

Materials

Time allowed: 1 hour 15 minutes

For this paper you must have:

- a ruler
- a calculator
- the periodic table (see page 170).

Instructions

- Answer **all** questions in the spaces provided.
- Do all rough work on the page. Cross through any work you do not want to be marked.

Information

- There are **70** marks available on this paper.
- The marks for each question are shown in brackets.
- You are expected to use a calculator where appropriate.
- You are reminded of the need for good English and clear presentation in your answers.
- When answering questions 03 and 05.2 you need to make sure that your answer:
  - is clear, logical, sensibly structured
  - fully meets the requirements of the question
  - shows that each separate point or step supports the overall answer.

Advice

- In all calculations, show clearly how you work out your answer.

Chemistry Practice Exam Paper 2

01 This question is about atmospheric pollutants from fuels.

Methane (CH_4) is a hydrocarbon which is used as a fuel.

The equation shows the reaction for the complete combustion of methane:

$$CH_4 + 2O_2 \rightarrow CO_2 + XH_2O$$

01.1 **X** represents what number?

Tick **one** box.

1 ☐

2 ☐

3 ☐

4 ☐   **[1 mark]**

If there is a limited amount of oxygen then methane will undergo incomplete combustion.
One product is carbon monoxide.

01.2 Which other product may be produced in this reaction?

Tick **one** box.

carbon particles ☐

sulfur dioxide ☐

nitrogen oxide ☐

hydrogen ☐   **[1 mark]**

01.3 Give **two** reasons why it is difficult to detect the toxic gas carbon monoxide.

_____

_____

_____

                                                                        [2 marks]

01.4 When methane is burned at high temperatures, oxides of nitrogen are produced by the reaction between nitrogen and oxygen from the air.

Which equation correctly shows the production of nitrogen dioxide?

Tick **one** box.

$N + O_2 \rightarrow NO_2$ ☐

$2N_2 + O_2 \rightarrow 2N_2O$ ☐

$N + O \rightarrow NO$ ☐

$N_2 + 2O_2 \rightarrow 2NO_2$ ☐

[1 mark]

01.5 Increased amounts of pollutants in the air cause problems.

Draw **one** line from each pollutant to the problem it causes.

Pollutant		Problem

sulfur dioxide		global warming
		global dimming
		acid rain
carbon particles		destruction of ozone layer

[2 marks]

Turn over for the next question

02   Cracking can be carried out in the laboratory using the apparatus shown in **Figure 1**.

Figure 1

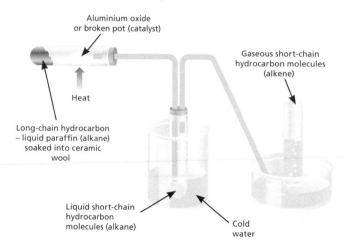

Aluminium oxide
or broken pot (catalyst)

Gaseous short-chain
hydrocarbon molecules
(alkene)

Heat

Long-chain hydrocarbon
– liquid paraffin (alkane)
soaked into ceramic
wool

Liquid short-chain
hydrocarbon
molecules (alkane)

Cold
water

One example of a cracking reaction is:

$C_{10}H_{22} \rightarrow C_7H_{16} + M$

02.1 State the formula and name of **M**.

Formula: ....

Name: ....   **[2 marks]**

02.2 Describe a test to show that the gas produced is an alkene.

Give the expected result of the test.

....

....

....   **[2 marks]**

An alkene called ethene is used to make a polymer called poly(ethene).

There are two types of poly(ethene): low density (LD) and high density (HD).

Table 1 shows the properties of the two types of poly(ethene).

Table 1

Property	LD Poly(ethene)	HD Poly(ethene)
Density	0.91–0.94 g/cm³	0.95–0.97 g/cm³
Flexibility	High	Low
Strength	Low	High

02.3 A manufacturer wants to make plastic buckets.

Suggest what type of poly(ethene) they should use.
Give a reason for your choice.

_____

_____

_____ [2 marks]

Turn over for the next question

03   An increase in average global temperature is a cause of climate change.

Explain the effects of global climate change on the environment, humans and wildlife.

....

....

....

....

....

....

....

....

....

....

....

....

....

.... **[6 marks]**

04 Humans need clean drinking water.

Water that is safe to drink is called potable water.

04.1 What is the name of the process that produces potable water from seawater?

.... [1 mark]

04.2 Potable water cannot be described as pure.

Explain why.

....

....

.... [2 marks]

04.3 A student is given the following equipment:

A beaker of water, a Bunsen burner, a tripod and gauze, and a thermometer.

Describe how the student can use the equipment to test if the water sample is pure.

You should state what they will observe if the water is pure in your answer.

....

....

.... [2 marks]

04.4 Name **one** sterilising agent used to kill microorganisms in water.

.... [1 mark]

Question 4 continues on the next page

04.5 Fluoride is sometimes added to water.

A sample of drinking water contains 1.35 mg of fluoride per dm^3 of water.

Calculate the mass of fluoride in 250 cm^3 of water.

Give your answer to 2 decimal places.

1000 cm^3 = 1 dm^3

Mass = _____ mg   **[3 marks]**

05 Steel and aluminium are used to make drinks cans.
They are both limited resources.

05.1 Explain what is meant by 'limited resources'.

....

.... **[1 mark]**

05.2 The metal from drinks cans can be recycled.

The following steps are used:
1. The cans are collected and sorted.
2. They are melted down and then cooled to form blocks of metal.
3. The blocks are rolled into thin sheets, which can be used to make new products.

Evaluate the use of recycling metal cans as a method of reducing the use of limited resources.

....

....

....

....

....

....

.... **[5 marks]**

06 Magnesium reacts with dilute hydrochloric acid:

$Mg(s) + 2HCl(aq) \rightarrow MgCl_2(aq) + H_2(g)$

A student investigated how the volume of hydrogen produced over time changes when magnesium is reacted with two different concentrations of dilute hydrochloric acid.

This is the method used:

1. Measure 20 cm³ of 0.5 mol/dm³ hydrochloric acid using a measuring cylinder.
2. Pour the acid into the conical flask.
3. Add 2 g of magnesium strip.
4. Place the bung into the flask.
5. Measure the volume of gas every 30 seconds until the reaction is complete.
6. Repeat using 1 mol/dm³ hydrochloric acid.

The student reached the end of **Step 2** and set up the apparatus as shown in **Figure 2**.

Figure 2

06.1 Identify what the student should do before continuing with the method.
Describe what could happen if the student continued without making any changes.
Explain how this would affect the results.

_____

_____

_____

_____

[3 marks]

Question 6 continues on the next page

The student corrected the error.

Their results are shown in **Table 2**.

Table 2

Time in s	Total volume of hydrogen in cm³	
	0.5 mol/dm³ acid	1 mol/dm³ acid
0	0.0	0.0
30	8.2	14.1
60	14.4	25.5
90	20.0	33.6
120	25.1	36.8
150	29.3	37.6
180	33.7	38.0
210	36.2	38.0
240	37.8	38.0
270	38.0	38.0

06.2 On **Figure 3**:
- Plot both sets of results on the grid.
- Draw two lines of best fit.

Figure 3

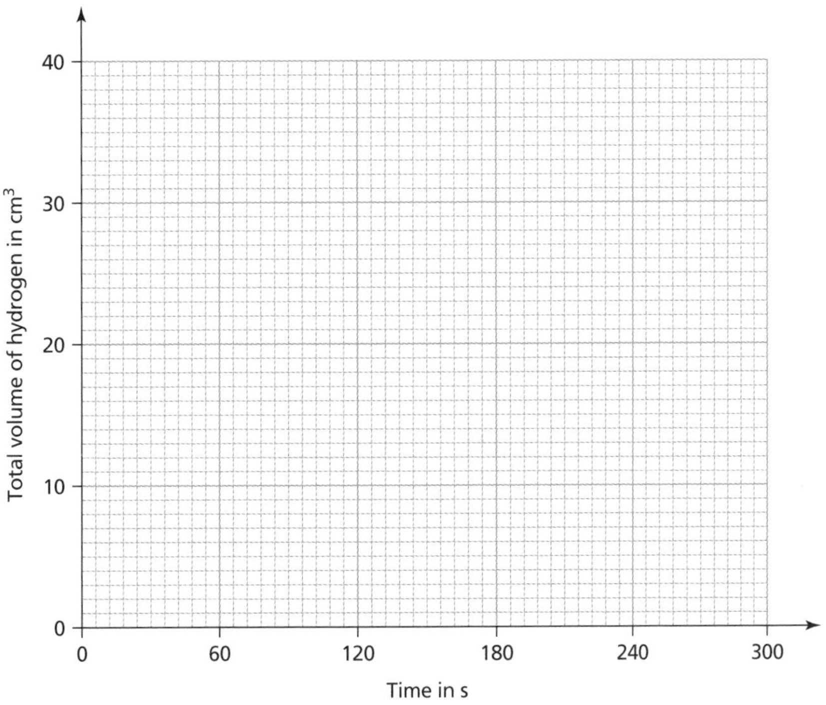

Time in s

[4 marks]

06.3 How does the concentration of acid affect the rate of the reaction?

....

.... **[1 mark]**

06.4 Explain why, in terms of particles, the concentration of acid affects the rate of the reaction.

....

....

....

....

.... **[3 marks]**

The student decided to research how the temperature of the acid affected the rate of reaction.

Figure 4 shows a graph the student found in a text book.

Figure 4

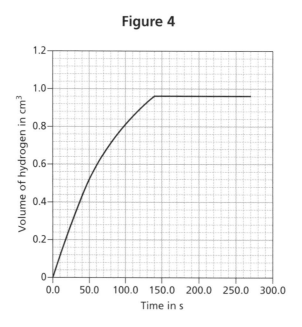

06.5 State the time at which the reaction finishes.

Answer: .... s **[1 mark]**

Question 6 continues on the next page

06.6 Use **Figure 4** to calculate the rate of the reaction at 50 seconds.

Give your answer to one significant figure.

Give the unit.

Rate of reaction = _____ **[6 marks]**

07 This question is about how the amounts of the different gases in the atmosphere have changed.

Figure 5 shows how the levels of oxygen in the atmosphere have changed since the Earth was formed.

Figure 5

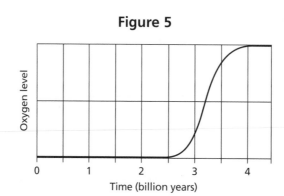

07.1 Use the graph to state when oxygen first started to be produced.

Answer: _____ billion years **[1 mark]**

07.2 The air today is approximately one-fifth oxygen.

Calculate the approximate volume of oxygen in 200 cm³ of air.

Answer: _____ cm³ **[1 mark]**

07.3 Explain how the amount of oxygen in the air increased to the amount found today.

_____

_____

_____

_____

_____

[3 marks]

Question 7 continues on the next page

Another gas found in the air today is carbon dioxide.

07.4 Describe how carbon dioxide helps to maintain temperatures on Earth.

[3 marks]

07.5 In what way has the amount of carbon dioxide in the atmosphere changed over the last 100 years?

[1 mark]

07.6 Describe **one** way in which human activity has brought about this change to the amount of carbon dioxide in the atmosphere today.

[2 marks]

08   Paper chromatography can also be used to identify substances.

Figure 6 shows the results from chromatography carried out on a mixture.

Figure 6

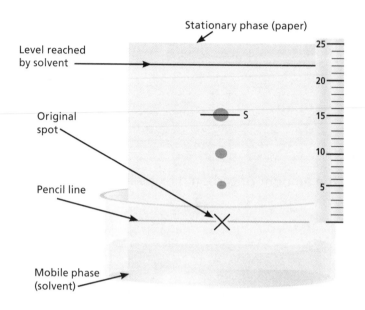

08.1 How many substances are present in the mixture?

Answer: .... [1 mark]

08.2 To identify a substance, its R_f value must be calculated.

Calculate the R_f value of **S**.
Give your answer to 2 decimal places.

Answer: .... [3 marks]

08.3 Explain how paper chromatography separates the substances in a mixture.

....

....

....
[3 marks]

END OF QUESTIONS

Collins

GCSE
COMBINED SCIENCE

H

Physics: Paper 1 Higher Tier

Materials

Time allowed: 1 hour 15 minutes

For this paper you must have:

- a ruler
- a calculator
- the Physics Equation Sheet (page 169).

Instructions

- Answer **all** questions in the spaces provided.
- Do all rough work on the page. Cross through any work you do not want to be marked.

Information

- There are **70** marks available on this paper.
- The marks for each question are shown in brackets [ ].
- You are expected to use a calculator where appropriate.
- You are reminded of the need for good English and clear presentation in your answers.
- When answering questions 01.2 and 04.5 you need to make sure that your answer:
  – is clear, logical, sensibly structured
  – fully meets the requirements of the question
  – shows that each separate point or step supports the overall answer.

Advice

- In all calculations, show clearly how you work out your answer.

01   **Figure 1** shows what happens to each 100 joules of energy from coal that is burned in a power station.

Figure 1

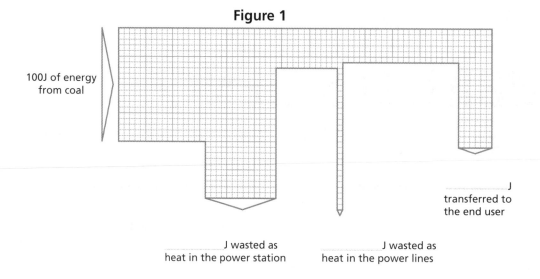

100J of energy from coal

.... J transferred to the end user

.... J wasted as heat in the power station

.... J wasted as heat in the power lines

01.1   Add the missing figures to the diagram.   **[3 marks]**

01.2   For the same cost, the electricity company could:
- install new power lines that only waste half as much energy as the old ones
  OR
- use a quarter of the heat wasted at the power station to heat schools in a nearby town.

Which of these two things do you think they should do?
Give a reason for your answer.

....

....

....

....

....

....   **[4 marks]**

01.3   Calculate the efficiency of the coal powered station in **Figure 1**.

Efficiency = ... %   **[1 mark]**

02   A gas burner is used to heat some water in a pan.
      By the time the water starts to boil:
      - 60% of the energy released has been transferred to the water
      - 20% of the energy released has been transferred to the surrounding air
      - 13% of the energy released has been transferred to the pan
      - 7% of the energy released has been transferred to the gas burner itself.

   02.1   Some of the energy released by the burning gas is wasted.

          What happens to this wasted energy?

          _____

          _____ **[2 marks]**

   02.2   What percentage of the energy from the gas is wasted?

                              Percentage = _____ %   **[1 mark]**

   02.3   How efficient is the gas burner at heating water?

                              Efficiency = _____ %   **[1 mark]**

Turn over for the next question

03 A book weighs 6 newtons.

A librarian picks up the book from the ground and puts it on a shelf that is 2 metres high.

03.1 Calculate the work done on the book.

Work done = _____ J **[2 marks]**

03.2 The next person to take the book from the shelf accidentally drops it.
The book falls 2 m to the ground.

Calculate how much gravitational energy it loses as it falls.
The gravitational field strength is = 10 N/kg.

Answer = _____ J **[2 marks]**

03.3 All of the book's gravitational energy is converted to kinetic energy when it falls.

Calculate the velocity with which the book hits the floor.

Velocity = _____ m/s **[3 marks]**

04   Electricity can be produced from a number of different energy resources.

04.1   Complete **Table 1** to show the energy transfers that occur for different resources.

Table 1

Device	Energy resource	Useful energy transfer from resource	
Coal-fired power station	Coal	chemical $\longrightarrow$	electrical
Hydroelectric power station	Stored water	$\longrightarrow$	electrical
Solar panel	Sun	$\longrightarrow$	electrical
Wind turbine	Wind	$\longrightarrow$	electrical
Gas-fired power station	Gas	$\longrightarrow$	electrical

[4 marks]

04.2   State which of the five energy resources in **Table 1** are not renewable.

.... **[1 mark]**

04.3   Give another non-renewable energy resource.

.... **[1 mark]**

04.4   Give a renewable energy source **not** listed in **Table 1**.

.... **[1 mark]**

Question 4 continues on the next page

04.5 State and explain the advantages and disadvantages of using nuclear power
stations to produce electricity.

[4 marks]

05 A student carries out an experiment to investigate the current through component **X**.
A circuit is set up as shown in **Figure 2**.
The current is measured when different voltages are applied across component **X**.

Figure 2

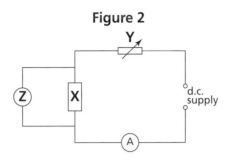

05.1 Name the components labeled **Y** and **Z** in the circuit.

Y = _____ [1 mark]

Z = _____ [1 mark]

05.2 What is the role of component **Y** in the circuit?

_____ [1 mark]

Table 2 shows the measurements obtained in this experiment.

Table 2

Voltage (V)	−0.6	−0.4	−0.2	0	0.2	0.4	0.6	0.8
Current (mA)	0	0	0	0	0	50	100	150

05.3 Name the independent variable in this experiment.

Independent variable = _____ [1 mark]

05.4 Name the dependent variable in this experiment.

Dependent variable = _____ [1 mark]

Question 5 continues on the next page

05.5 Plot a graph on the axes in **Figure 3** using the data from **Table 2**.

Figure 3

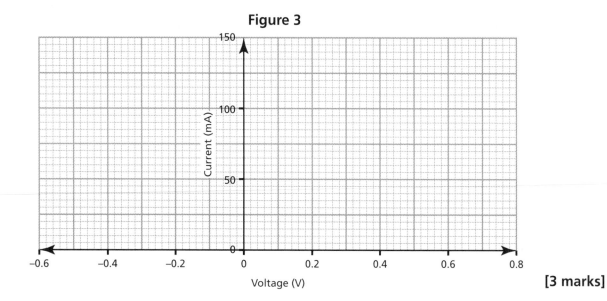

Voltage (V)

[3 marks]

05.6 The student looks at their measurements and decides that there are no anomalous results.

Are they correct?
You must explain your answer.

....

.... [1 mark]

05.7 Use the shape of the graph to name component **X**.

Component **X** = ... [1 mark]

06 There are many isotopes of the element strontium (Sr).

06.1 What do the nuclei of different strontium isotopes have in common?

.... **[1 mark]**

06.2 When the nucleus of a strontium-90 atom decays, it emits radiation and changes into a nucleus of yttrium-90

$$^{90}_{38}Sr \rightarrow ^{90}_{39}Y + Radiation$$

What type of decay is this?

Answer .... **[1 mark]**

06.3 Give a reason for your answer to **06.2**.

.... **[1 mark]**

Strontium-90 has a half-life of 30 years.

06.4 What is meant by the term 'half-life'?

.... **[1 mark]**

06.5 After formation in the nuclear reactor, strontium-90 is stored as radioactive waste.

For how many years does strontium-90 have to be stored before its radioactivity has fallen to $\frac{1}{8}$ of its original level?

Answer .... **[2 marks]**

Turn over for the next question

07 A car that is moving has kinetic energy.
The faster a car goes, the more kinetic energy it has.

07.1 The kinetic energy of a car was 472 500 J when travelling at 30 m/s.

Calculate the total mass of the car.
Give the unit.

Mass = ... **[4 marks]**

There is a government road safety campaign to reduce the speed at which people drive in residential areas.
It uses the slogan 'Kill your speed, not a child'.
The scientific reason for this is that kinetic energy is transferred from the vehicle to the person it knocks down.

07.2 A bus and car are travelling at the same speed.
The bus is likely to cause more harm to a person who is knocked down than the van would.

Explain why.

....

....

....

[2 marks]

07.3 A car and its passengers have a total mass of 1200 kg.
The car is travelling at 8 m/s.

Calculate the increase in kinetic energy when the car increases its speed to 14 m/s.

Increase .... J   [3 marks]

07.4 Explain why the increase in kinetic energy is much greater than the increase in speed.

....

.... [1 mark]

Turn over for the next question

08 In order to jump over the bar, a high jumper must raise his mass above the ground by 1.25 m. The high jumper has a mass of 65 kg.

The gravitational field strength is 10 N/kg.

08.1 The high jumper just clears the bar.

Calculate the gain in his gravitational potential energy.

Gain = _____ J   **[2 marks]**

08.2 Calculate the minimum vertical speed the high jumper must reach in order to jump over the bar.

Use your answer to **08.1** and the formula for kinetic energy.

Minimum vertical speed = _____ m/s   **[3 marks]**

09 The circuit diagram in **Figure 4** shows a circuit used to supply electricity for car headlights.

Figure 4

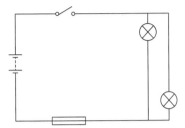

The current through the filament of one car headlight is 2 A.
The potential difference supplied by the battery is 12 V.

09.1 What is the potential difference across each headlight?

Potential difference = ... V   **[1 mark]**

09.2 Work out the total current through the battery.

Current = ... A   **[1 mark]**

09.3 Calculate the resistance of each headlight filament when in use.

Resistance = ... Ω   **[2 marks]**

Question 9 continues on the next page

09.4 How does the total resistance of the circuit compare to the resistance of each individual bulb?

_____ [1 mark]

09.5 Calculate the power supplied to each of the two headlights of the car.

Power = _____ W   [2 marks]

09.6 The fully charged car battery can deliver 96 kJ of energy at 12 V.

How long can the battery keep both the headlights fully on?

Length of time = _____ s   [2 marks]

END OF QUESTIONS

Collins

GCSE
COMBINED SCIENCE

H

Physics: Paper 2 Higher Tier

Materials

Time allowed: 1 hour 15 minutes

For this paper you must have:

- a ruler
- a calculator
- a protractor
- the Physics Equation Sheet (page 169).

Instructions

- Answer **all** questions in the spaces provided.
- Do all rough work on the page. Cross through any work you do not want to be marked.

Information

- There are **70** marks available on this paper.
- The marks for each question are shown in brackets [ ].
- You are expected to use a calculator where appropriate.
- You are reminded of the need for good English and clear presentation in your answers.
- When answering questions 04.1 and 06.5 you need to make sure that your answer:
  - is clear, logical, sensibly structured
  - fully meets the requirements of the question
  - shows that each separate point or step supports the overall answer.

Advice

- In all calculations, show clearly how you work out your answer.

01   A student used a lever system to investigate how the force of attraction between a coil and an iron rocker varied with the current in the coil.
She supported a coil vertically and connected it to an electrical circuit as shown in **Figure 1**.
The weight of the iron rocker is negligible.

Figure 1

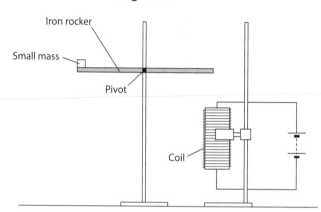

01.1 Why is it important that the rocker in this experiment is made of iron?

....   **[1 mark]**

01.2 The student put a small mass on the end of the rocker and adjusted the current in the coil until the rocker balanced.

To keep the rocker balanced, how will the current through the coil need to change as the size of the mass is increased?

....   **[1 mark]**

01.3 Explain your answer to **01.2**.

....

....

....   **[2 marks]**

01.4 A second student set up the same experiment and put an iron core inside the coil.

How will this affect the size of the mass that can be balanced?
You must explain you answer.

_____

_____

_____ **[2 marks]**

Turn over for the next question

02   A group of students investigate circular motion.
They swing a bung attached to a string around in circle.
The string is attached to a force meter, which measures
the centripetal force (the force that keeps an object on
a circular path).

Figure 2

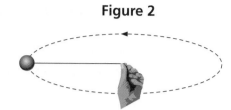

The students record how the reading on the force meter
changes to determine how the force affects the speed of the orbit.

02.1   In which direction does the centripetal force act on the rubber bung?

          [1 mark]

02.2   In this experiment, what provides the centripetal force?

          [1 mark]

02.3   One student swung the rubber bung around in a circle at constant speed.
A second student timed how long it took the rubber bung to complete 10 rotations.

Give **two** variables that are important to control in this experiment.

          [2 marks]

02.4   The Moon orbits the Earth in a circular path.

direction	resistance	speed	velocity

Use words from the box to complete the sentences.
You may use each word once, more than once or not at all.

The Moon's _____ is constant but its _____ changes.

This is because its _____ changes.   **[3 marks]**

02.5   What force provides the centripetal force needed to keep the Moon in its orbit
around the Earth?

          [1 mark]

03   Radio waves, ultraviolet waves, visible light and X-rays are all types of electromagnetic radiation.

03.1   Choose wavelengths from the list below to complete **Table 1**.

3×10^{-8} m 1×10^{-11} m 5×10^{-7} m   1500 m

Table 1

Type of Radiation	Wavelength (m)
Radio waves	
Ultraviolet waves	
Visible light	
X-rays	

[3 marks]

03.2   Give **two** properties that are common to all the waves in **Table 1**.

....

....   [2 marks]

03.3   State the name and a use of **one** type of electromagnetic wave **not** listed in **Table 1**.

....

....   [2 marks]

03.4   Radio waves can be used to control remote control cars.

Calculate the frequency of radio waves of wavelength 300m.
(The velocity of electromagnetic waves is 3×10^8 m/s.)

Frequency = ...   Hz   [3 marks]

Question 3 continues on the next page

The graph in **Figure 3** shows the speed of a remote-controlled vehicle during a race.

Figure 3

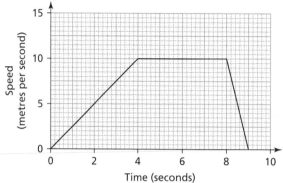

03.5 Calculate the acceleration during the first four seconds.

Acceleration = ... m/s²   **[3 marks]**

03.6 What is the maximum speed reached by the vehicle?

Maximum speed = ... m/s   **[1 mark]**

03.7 How far does the vehicle travel between 4 and 8 seconds?

Distance = ... m   **[2 marks]**

04 The distance–time graph in **Figure 4** represents the motion of a car during a race.

Figure 4

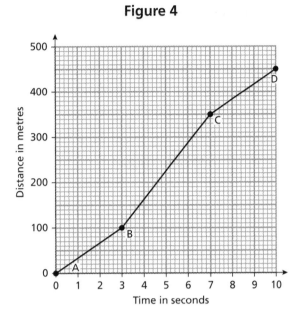

04.1 Describe the motion of the car between points **A** and **D**.
You should calculate the maximum speed reached by the car and include this in your answer.

_____

_____

_____

_____

_____

_____

_____ **[6 marks]**

Question 4 continues on the next page

04.2   At the start of the race, the car accelerates from rest to a speed of 30 m/s in 6 seconds.

Calculate the acceleration of the car.

Acceleration = ... m/s²   **[3 marks]**

05 A car of mass 1200 kg has an engine thrust of 3500 N and experiences a resistive force of 2000 N.

05.1 Calculate the acceleration of the car.

Acceleration = _____ m/s² **[4 marks]**

05.2 Explain why the car reaches a top speed even though the thrust force remains constant at 3500 N.

_____

_____

_____ **[2 marks]**

The driver of the car notices a hazard.
They apply the brakes to come to a complete stop in a certain distance.
This stopping distance is made up of the thinking distance and the braking distance.

05.3 What is meant by the term 'thinking distance'?

_____

_____ **[1 mark]**

05.4 State **two** factors that affect thinking distance.

_____

_____

_____

_____ **[2 marks]**

Question 5 continues on the next page

05.5 The braking distance of a car depends on the speed of the car and the braking force applied.

State **one** other factor that affects the braking distance.

_____

_____ **[1 mark]**

05.6 During braking, the temperatures of the brakes increase.

Explain why.

_____

_____

_____

_____ **[2 marks]**

06 **Figure 5** shows two waves.

Figure 5

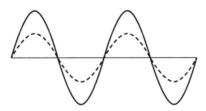

06.1 Name **one** wave quantity that is the same for both waves.

.... [1 mark]

06.2 Name **one** wave quantity that is different for the two waves.

.... [1 mark]

06.3 The waves shown in **Figure 5** are transverse waves.

Which of the following types of wave is **not** a transverse wave?
Draw a ring around the correct answer.

gamma rays   **sound**   **visible light**   [1 mark]

06.4 A student studies waves in a ripple tank.
Every second, eight waves pass a fixed point in the tank.
The waves have a wavelength of 0.015 m.

Calculate the speed of the water waves.

Wave speed = ... m/s   [3 marks]

Question 6 continues on the next page

06.5 Measuring the wavelength of moving water waves is difficult.
To improve accuracy, the student decides to use a stroboscope and a ruler.

Describe the procedure that the student should use and outline **one** hazard they need to be aware of.

_____

_____

_____

_____

_____

[4 marks]

07 **Figure 6** shows two lines of the magnetic field pattern around a current-carrying wire.

Figure 6

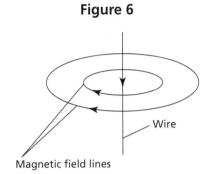

07.1 The direction of the current is reversed.

How does this affect the lines in the magnetic field pattern?

.... **[1 mark]**

07.2 The size of the current through the wire is increased.

How does this affect the lines in the magnetic field pattern?

.... **[1 mark]**

Question 7 continues on the next page

07.3 Fleming's left hand rule can be used to identify the direction of a force acting on a current-carrying wire in a magnetic field.

Figure 7

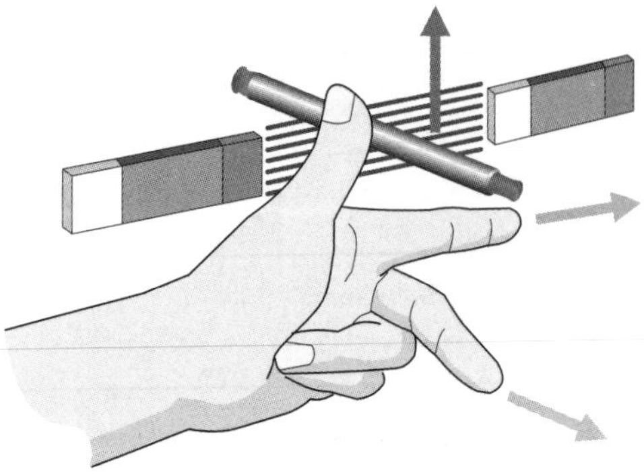

Complete the table below to show what is indicated by each part of the hand.

Part of hand	Indicates the direction of:
First finger	
Second finger	
Thumb	

[3 marks]

Figure 8 illustrates a pair of magnets.
The arrow shows the direction of the current in a wire passing between the magnets.

Figure 8

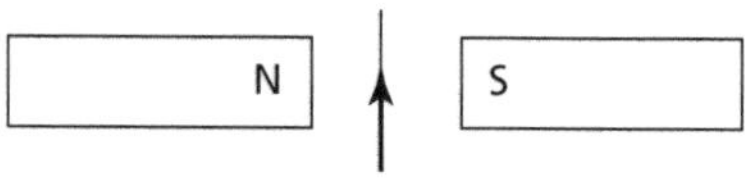

07.4 In which direction does the force on the wire act?

[1 mark]

07.5 Give **three** changes that would **decrease** the force acting on the wire.

[3 marks]

END OF QUESTIONS

Physics Equation Sheet

1	(final velocity)² – (initial velocity)² = 2 × acceleration × distance	$v^2 - u^2 = 2as$
2	elastic potential energy = 0.5 × spring constant × (extension)²	$E_e = \frac{1}{2}ke^2$
3	change in thermal energy = mass × specific heat capacity × temperature change	$\Delta E = mc\Delta\theta$
4	period = $\dfrac{1}{\text{frequency}}$	
5	**force on a conductor (at right-angles to a magnetic field) carrying a current = magnetic flux density × current × length**	$F = BIl$
6	thermal energy for a change of state = mass × specific latent heat	$E = mL$
7	**potential difference across primary coil × current in primary coil = potential difference across secondary coil × current in secondary coil**	$V_s I_s = V_p I_p$

The Periodic Table

Key

relative atomic mass
atomic symbol
name
atomic (proton) number

1	2												3	4	5	6	7	0
						1 **H** hydrogen 1												4 **He** helium 2
7 **Li** lithium 3	9 **Be** beryllium 4												11 **B** boron 5	12 **C** carbon 6	14 **N** nitrogen 7	16 **O** oxygen 8	19 **F** fluorine 9	20 **Ne** neon 10
23 **Na** sodium 11	24 **Mg** magnesium 12												27 **Al** aluminum 13	28 **Si** silicon 14	31 **P** phosphorus 15	32 **S** sulfur 16	35.5 **Cl** chlorine 17	40 **Ar** argon 18
39 **K** potassium 19	40 **Ca** calcium 20	45 **Sc** scandium 21	48 **Ti** titanium 22	51 **V** vanadium 23	52 **Cr** chromium 24	55 **Mn** manganese 25	56 **Fe** iron 26	59 **Co** cobalt 27	59 **Ni** nickel 28	63.5 **Cu** copper 29	65 **Zn** zinc 30		70 **Ga** gallium 31	73 **Ge** germanium 32	75 **As** arsenic 33	79 **Se** selenium 34	80 **Br** bromine 35	84 **Kr** krypton 36
85 **Rb** rubidium 37	88 **Sr** strontium 38	89 **Y** yttrium 39	91 **Zr** zirconium 40	93 **Nb** niobium 41	96 **Mo** molybdenum 42	[98] **Tc** technetium 43	101 **Ru** ruthenium 44	103 **Rh** rhodium 45	106 **Pd** palladium 46	108 **Ag** silver 47	112 **Cd** cadmium 48		115 **In** indium 49	119 **Sn** tin 50	122 **Sb** antimony 51	128 **Te** tellurium 52	127 **I** iodine 53	131 **Xe** xenon 54
133 **Cs** caesium 55	137 **Ba** barium 56	139 **La*** lanthanum 57	178 **Hf** hafnium 72	181 **Ta** tantalum 73	184 **W** tungsten 74	186 **Re** rhenium 75	190 **Os** osmium 76	192 **Ir** iridium 77	195 **Pt** platinum 78	197 **Au** gold 79	201 **Hg** mercury 80		204 **Tl** thallium 81	207 **Pb** lead 82	209 **Bi** bismuth 83	[209] **Po** polonium 84	[210] **At** astatine 85	[222] **Rn** radon 86
[223] **Fr** francium 87	[226] **Ra** radium 88	[227] **Ac*** actinium 89	[261] **Rf** rutherfordium 104	[262] **Db** dubnium 105	[266] **Sg** seaborgium 106	[264] **Bh** bohrium 107	[277] **Hs** hassium 108	[268] **Mt** meitnerium 109	[271] **Ds** darmstadtium 110	[272] **Rg** roentgenium 111	[285] **Cn** copernicium 112		[286] **Uut** ununtrium 113	[289] **Fl** flerovium 114	[289] **Uup** ununpentium 115	[293] **Lv** livermorium 116	[294] **Uus** ununseptium 117	[294] **Uuo** ununoctium 118

* The Lanthanides (atomic numbers 58–71) and the Actinides (atomic numbers 90–103) have been omitted.

Relative atomic masses for **Cu** and **Cl** have not been rounded to the nearest whole number.

Answers

Topic-Based Questions

Page 6 Cell Structure

1. a) B   [1]
   b) A   [1]
   c) C   [1]
2. a) B [1]; it has a nucleus / chloroplasts [1]
   b) Eukaryotic cells are much larger than prokaryotic   [1]
   c) Using flagella [1]; which whip around [1]

Page 7 Investigating Cells

1. a) To act as a stain [1]; because some of the structures are transparent [1]
   b) i) 30 micrometres   [1]
      ii) $\frac{20}{0.03}$ [1]; = 667, magnified 667 times (×667) [1]
2. a) 0.11mm   [1]
   b) 10 micrometres   [1]

Page 8 Cell Division

1. a) Mitosis   [1]
   b) P, Q, (S), R [2] (1 mark for one in the correct place)
2. a) Stem cells can divide to produce different types of cells [1]; these cells can replace lost or defective cells or be grown into tissues [1]
   b) **Any one of:** the stem cells may come from cloned embryos [1]; embryos may be destroyed in the process [1]; concerned that the long-term effects are not known [1]

Page 9 Transport In and Out of Cells

1. a) 6 × (10 × 10) = 600   [1]
   b) Diffusion   [1]
   c) Block C [1]; it has the largest surface area [1]; so there is more surface for the dye to diffuse across [1]
2. a) Osmosis always involves movement of water particles, whereas diffusion is movement of particles [1]; osmosis involves a cell membrane and diffusion does not [1]
   b) Active transport is movement of particles against a concentration gradient / from low to high concentration, rather than from high to low concentration as with diffusion [1]; active transport requires energy from respiration (diffusion does not require an input of energy from the cell) [1]

Page 10 Levels of Organisation

1. a) cells, tissues, organs, systems   [1]
   b) **Top to bottom:** organ [1]; cell [1]; organ [1]; tissue [1]
   c) Muscle cells are specialised for contraction.   [1]
2. Phloem cells join end to end [1]; the end walls have holes to let sugar through

[1]; xylem cells join end to end with end walls broken down [1]; to form hollow tubes / strengthened with lignin [1]

Page 11 Digestion

1. a) The rate of reaction increases to a peak / optimum [1]; and then decreases [1]
   b) 40°C   [1]
2. a) Pancreas / small intestine   [1]
   b) i) There are more than three fatty acids / it contains sugar / no glycerol is present   [1]
      ii) Olestra is a different shape to fats [1]; so it cannot fit into the enzyme's active site [1]

Page 12 Blood and the Circulation

1. 3, 1, 2, 4 [3] (2 marks for two correct; 1 mark for one correct)
2. a) Haemoglobin combines with oxygen at the lungs, when in high concentration [1]; and forms oxyhaemoglobin [1]; it releases oxygen at the tissues, as the concentration is low there, to go back to haemoglobin [1]

   haemoglobin + oxygen ⇌ oxyhaemoglobin

   b) Lower surface area to take up or release oxygen [1]; shape means they get stuck in capillaries causing pain [1]

Page 13 Non-Communicable Diseases

1. a) i) The heart stops beating (Accept an accurate explanation of why)   [1]
      ii) It restricts blood flow to the heart [1]; so not enough oxygen and glucose are supplied to cells in the heart muscle for respiration [1]
   b) Having an atheroma does not seem to depend on level of physical activity [1]; increased exercise reduced incidence of heart disease [1]; suggesting atheromas are not the only factor in causing heart disease [1]

Page 14 Transport in Plants

1. a) water [1]; transpiration [1]; windy [1]
   b) i) A smaller change in mass due to less transpiration [1]; because less water is taken up by the roots [1]
      ii) A smaller change in mass due to less transpiration [1]; as some of the stomata have been blocked [1]
2. **Any two of:** water is transported through the xylem, sugars are

transported in phloem [1]; water is moved upwards only, sugars move all over [1]; water movement does not need energy from respiration, sugars move by active transport [1]

Page 15 Pathogens and Disease

1. a) Four correctly draw lines [3] (2 marks for two lines; 1 mark for one line)
   rose black spot – fungus
   salmonella – bacterium
   measles – virus
   malaria – protozoan
   b) vector [1]; *Plasmodium* [1]; host [1]
2. a) **Any two of:** infected needles [1]; sexual activity [1]; across the placenta [1]
   b) The virus attacks white blood cells [1]; making the immune system less effective [1]
3. To prevent the spread of the disease [1]; to destroy the spores [1]

Page 16 Human Defences Against Disease

1. Nose: hairs trap particles [1]; skin: sebum kills bacteria [1]; stomach: acid kills microorganisms [1]
2. A pathogen has antigens on its surface [1]; lymphocytes make antibodies [1]; that attach to antigens on pathogens (with a specific fit) [1]; phagocytes ingest pathogens [1]

Page 17 Treating Diseases

1. a) HIV is a virus [1]; TB is a bacterial infection [1]
   b) Many TB bacteria are resistant [1]; and are not killed by antibiotics [1]
   c) **Any two of:** finish the dose [1]; only prescribe if necessary [1]; rotate antibiotics used [1]
   d) Use of a placebo [1]; patient does not know if they are having the placebo / real drug [1]; neither does the doctor [1]

Page 18 Photosynthesis

1. a) carbon dioxide + water [1]; ⟶ glucose + oxygen [1]
   b) i) **Any two of:** root tip [1]; shoot tip [1]; fruits [1]; seeds [1]; storage organs [1];
      ii) **Any two of:** starch [1]; for storage [1] OR lipids [1]; for storage [1] OR proteins [1]; for growth [1] OR cellulose [1]; for cell walls [1]
2. a) Some of the mass did come from the water [1]; but some also comes from carbon dioxide / photosynthesis [1]
   b) He showed that carbon dioxide was needed as well as water   [1]

Answers

Page 19 Respiration and Exercise

1. a) oxygen [1]; → carbon dioxide [1]; + water [1] (The two products can be given in any order)
   b) i) In the race / after training, lactic acid does not start to increase so early [1]; and does not increase as much [1]; lactic acid causes muscle cramps, and because there is less she can run more efficiently [1]
      ii) The lactic acid is sent to the liver [1]; where it is broken down [1]; using oxygen [1]

Page 20 Homeostasis and the Nervous System

1. a) Insulation [1]; increases the speed of the impulse [1]
   b) Transmitter molecules are released when an impulse reaches a synapse [1]; they diffuse across the synapse [1]; and stimulate an impulse in the next neurone [1]
   c) i) The drug mimics the neurotransmitter [1]; and stimulates an impulse in the next neurone [1]
      ii) Neurotransmitter molecules will not be broken down [1]; so they continue to stimulate the next neurone [1]

Page 21 Hormones and Homeostasis

1. a) Type 1 [1]
   b) Insulin [1]
   c) **Any two of:** blood glucose level too high [1]; glucose starts to pass out in urine [1]; coma and / or death [1]
   d) i) Type 2 [1]
      ii) Treat by modifying diet [1]; rather than by insulin injections [1]
2. Less water is reabsorbed back into the body [1]; urine is less concentrated / higher volume [1]; which could lead to dehydration [1]

Page 22 Hormones and Reproduction

1. a) Hormone A: Oestrogen [1]; inhibits FSH release / stimulates LH release / repairs lining of uterus [1]; Hormone B: progesterone [1]; maintains the lining of the uterus / inhibits FSH and LH [1]
   b) i) **X** marked just after peak of Hormone **A** [1]
      ii) **Y** marked at start or end of the cycle [1]
   c) i) Stimulates follicle development [1]; ready for ovulation [1]
      ii) Causes release of an egg [1]
2. The pill contains both oestrogen and progesterone [1]; it inhibits FSH [1]; so that eggs do not develop / no ovulation [1]

Page 23 Sexual and Asexual Reproduction

1. a) It produces runners [1]; that touch the ground and root, growing into new plants [1]
   b) **Any two of:** it produces flowers [1]; pollen is transferred from plant to plant [1]; it makes seeds [1]
2. Egg cell: meiosis [1]; eye cell: 38 and mitosis [1]; sperm cell: 19 [1]; leg cell: 38 [1]

Page 24 Patterns of Inheritance

1. a) i) Sachin and Rose [1]
      ii) Sara, Tia and Rohit [1]
   b) Sachin: gg (genotype) [1]; male with Gaucher disease (phenotype) [1]; Tia: female with normal enzyme (phenotype) [1]
   c) Probability: 50% / $\frac{1}{2}$ / 1 : 1 [1];

	G	g	
g	Gg	gg	[1];
g	Gg	gg	[1]

[1];

   d) **Any two of:** She might prefer not know until it is born [1]; to avoid having to decide whether to have a termination [1]; the test may increase the risk of miscarriage [1] (Accept any other sensible reason)

Page 25 Variation and Evolution

1. Genetic: Bill and Ben have brown eyes [1]; Environment: Ben has a scar [1]; Genetics and Environment: Bill is 160cm tall [1]; Ben's body mass is 60kg [1]
2. a) Darken / turn black [1]
   b) The dark moths are better camouflaged [1]; so fewer are eaten by birds [1]

Page 26 Manipulating Genes

1. Selective breeding [1]; genetic engineering [1]
2. a) (D), A, B, C [2] (1 mark for one in the correct place)
   b) The process uses hormones to cut DNA. [1]; The insulin molecules made are a mixture of human and bacterial insulin. [1]

Page 27 Classification

1. a) Organisms that can mate with each other [1]; to produce fertile offspring [1]
   b) i) Bobcat [1]; ocelot [1]
      ii) They are both in the same genus [1]; *Felis* [1]
2. a) i) 160 million years ago [1]
      ii) Archaeopteryx [1]
   b) **Any two of:** new disease [1]; new predators [1]; more successful competitors [1]; catastrophic event, e.g. meteor [1]

Page 28 Ecosystems

1. a) i) To spread their weight / stop them sinking in the sand [1]
      ii) To shade them from the sun / store of food or fat / produce water from respiration (by breaking down the fat store) [1]
   b) population [1]; habitat [1]; community [1]; competed [1]; biotic [1]
   c) i) Use a quadrat [1]; placed at random [1]; count cacti present in quadrat [1]; repeat many times [1]; calculate the mean number per m² [1]; multiply results up by the area [1]
      ii) Camels move around [1]; so may count them more than once / not at all [1]

Page 29 Cycles and Feeding Relationships

1. A = Photosynthesis [1];
   B = Respiration [1];
   C = Decomposition [1];
   D = Combustion / burning [1];
   E = Eating / feeding [1]
2. a) Predator–prey graph [1]
   b) Rabbit numbers increase, so there is more food available for the foxes [1]; more foxes survive and numbers increase [1]; this means more rabbits are eaten, so the fox numbers drop again (as less food is available) [1]

Page 30 Disrupting Ecosystems

1. a) There is no pollution / no industry / no cars [1]
   b) **Any two of:** in summer, there is more sunlight [1]; so more photosynthesis [1]; and more carbon dioxide taken in by plants [1]
   c) Greenhouse effect / greenhouse gas [1]; allows heat from the Sun through the atmosphere [1]; but absorbs / traps energy reflected back from the Earth's surface, leading to a rise in temperature [1]
2. a) Frog [1]
   b) When some fossil fuels are burned [1]; acidic gases are released, e.g. sulfur dioxide gas [1]; which dissolve in water in the atmosphere, which then falls as acid rain [1]

Page 31 Atoms, Elements, Compounds and Mixtures

1. a) $2Cu + O_2 \rightarrow 2CuO$ [1]
   b) 18.9 – 15.6 = 3.3g [1]
   c) 0.1g [1]

The resolution of the balance is the degree of accuracy. In this case, the measurements are given to the nearest 0.1g.

2. a) Place the salt solution in the round bottomed flask [1]; heat the solution [1]; and collect the liquid that distils at 100°C [1]
   b) **Any one of:** Burns from the Bunsen burner / hot water / steam [1]; equipment may crack so risk of cuts [1]

Page 32 Atoms and the Periodic Table

1. a) Thomson showed that the atom could be divided up into simpler substances [1]; the old model was no longer correct [1]
   b) Electron [1]
2. a) Based on Thomson's model, the alpha particles would have all gone through [1] OR There would have been no deflection [1] (Accept: It went against Thomson's model)
   b) To reduce the effects of errors [1]; to check repeatability [1] (Accept: To make sure they were correct / accurate [1])
   c) **Any three of:** small nucleus [1]; nucleus in the centre of the atom [1]; nucleus has a positive charge [1]; electrons in orbit around the nucleus [1] (Accept a labelled diagram)

Page 33 The Periodic Table

1. C [1]

The number of electrons in the outer shell is the same as the group that the element is in. The number of electron shells is the same as the row (period) that the element is in.

2. a) **X** [1]; because it does not conduct electricity [1]
   b) **W** [1]; because it is a liquid at room temperature [1]; and it conducts electricity [1]
   c) (lithium) hydroxide [1]; hydrogen [1]

Page 34 States of Matter

1. a) g [1]; s [1]

Gas = g, solid = s, liquid = l, aqueous / dissolved in water = aq

   b) 650°C [1]
   c) −183°C [1]
   d) **Any six of:** Oxygen has a low boiling point [1]; magnesium oxide has a high boiling point [1]; oxygen has weak forces of attraction between its particles / molecules [1]; magnesium has strong forces between its ions / particles [1]; oxygen is a simple molecule [1]; magnesium oxide is an ionic compound [1]; only a small amount of energy is needed to boil oxygen / turn it from a liquid into a gas [1]; a lot of energy is needed to boil magnesium oxide / turn it from a liquid into a gas [1]

Page 35 Ionic Compounds

1. a) calcium chloride [1]; sodium carbonate [1]

Ionic compounds contain both metal and non-metal elements. Look at the periodic table if you are not sure whether an element is a metal or a non-metal. The metals are on the left side, the non-metals on the right.

   b) They cannot conduct electricity when solid because the ions cannot move [1]; they can conduct electricity when in solution because the ions are free to move about [1]; and carry the charge [1]
2. Electrons are transferred from magnesium to bromine [1]; the magnesium atom loses two electrons [1]; forming Mg^{2+} / 2+ ions [1]; the two bromine atoms each gain one electron [1]; forming Br^- / 1− ions [1]

Page 36 Covalent Compounds

1. The forces between bromine molecules are stronger [1]
2. a) A shared pair of electrons drawn between H and Br [1]; no additional hydrogen electrons and three non-bonding pairs shown on bromine [1] (second mark dependent on first)
   b) HBr [1]
3. High melting point – Strong covalent bonds between many carbon atoms [1]; Does not conduct electricity when molten – There are no charged particles that are free to move [1]

The covalent bonds between atoms are very strong. The force of attraction between molecules is called the intermolecular force and is much weaker.

Page 37 Metals and Special Materials

1. High melting point: strong force of attraction between positive ions and negative electrons / metallic bond is strong [1]; so, a lot of energy is needed to break metallic bonds (to melt metal) [1]; thermal conductivity: electrons are delocalised [1]; and are free to move through the metal and transfer energy [1]; malleability: the ions are arranged in layers / have a regular arrangement [1]; the layers are able to slide over each other easily [1]

Page 38 Conservation of Mass

1. a) $Zn(s) + 2H^+(aq) \rightarrow Zn^{2+}(aq) + H_2(g)$ [1]

In an ionic equation, only the ions that change (gain or lose electrons) are shown. They must be balanced.

   b) Gas is produced in the flask (increasing the pressure inside) [1]; the flask may crack / the bung may be forced out [1] (Accept: 'it is not safe' for 1 mark)
   c) The mass reading will decrease [1]; because the hydrogen particles (atoms) in the hydrochloric acid are rearranged [1]; to form hydrogen gas, which leaves the flask into the air [1]

Page 39 Amount of Substance

1. a) $2 \times 2 \times 2 = 8cm^3$ [1]
   b) mass = density × volume [1]; $10.49 \times 8 = 83.92g$ [1]
   c) moles = $\frac{mass}{A_r}$ [1]; $\frac{83.92}{108} = 0.78$ moles [1]
   d) 0.5mol of oxygen (O_2) [1]; 12g of carbon (C) [1]

Page 40 Reactivity of Metals

1. a) magnesium [1]; sodium [1]
   b) i) Carbon [1]
      ii) Zinc(II) oxide [1]
2. Oxidation: $Zn \rightarrow Zn^{2+} + 2e^-$ [1]; Reduction: $Cu^{2+} + 2e^- \rightarrow Cu$ [1]

Remember, OIL RIG: Oxidation Is Loss (of electrons); Reduction Is Gain (of electrons)

Page 41 The pH Scale and Salts

1. 4 [1]
2. a) Add excess copper(II) oxide to acid (accept alternatives, e.g. 'until no more will react') [1]; filter (to remove excess copper(II) oxide) [1]; heat filtrate to evaporate some water or heat to point of crystallisation [1]; leave to cool (so crystals form) [1]
   b) **Any one of:** wear apron [1]; use eye protection [1]; tie hair back [1]
   c) Copper(II) chloride [1]

Page 42 Electrolysis

1. a) Aluminium is more reactive than carbon [1]
   b) 2 [1]; 4 [1]
   c) It requires a lot of electricity to melt the aluminium oxide / keep it molten [1]; so aluminium oxide is dissolved in molten cryolite [1]; to reduce the melting point, so less electricity is used [1]

Page 43 Exothermic and Endothermic Reactions

1. a) It reduces the movement of heat to and from the surroundings [1]; which could affect the accuracy of the results [1] (Accept 'it is an insulator' for 1 mark)

Answers

b) **Any two of:** type of acid [1]; concentration of acid [1]; surface area of metal [1]; temperature of acid [1]; volume of acid [1]; mass of metal [1]

c) The more reactive the metal [1]; the more exothermic the reaction [1]

Page 44 Measuring Energy Changes

1. a) 366×2 [1]; $= 732$ [1]
   b) $436 + 193$ [1]; $= 629$ [1]
   c) Endothermic [1]; because the amount of energy needed to break the bonds in the reactants is greater than the energy released from forming new bonds [1]

Page 45 Rate of Reaction

1. a) Repeat using different concentrations of acid [1]; for example, $0.5mol/dm^3$, $1mol/dm^3$, $1.5mol/dm^3$, $2mol/dm^3$ [1] (Accept any other concentrations within a sensible range)
   b) **Any one of:** wear eye protection [1]; wear apron [1]; do not heat mixture over 50°C [1]
   c) The time taken for the cross to disappear will decrease as concentration of acid increases [1]; as the (acid) concentration increases so does the number of (acid) particles in a given volume [1]; so they collide more often with the sodium thiosulfate particles [1]; resulting in more successful collisions and a reaction taking place (sulfur forming) [1]

Page 46 Reversible Reactions

1. a) Reversible (reaction) [1]
   b) Forward reaction [1]; the backward reaction is endothermic [1]; because heat is needed to decompose the ammonium chloride [1]
   c) The forward and reverse reactions are taking place at the same rate. [1]; The amounts of reactants and products are constant. [1]
   d) More ammonium chloride will be produced [1]; until equilibrium is reached again [1]

Page 47 Alkanes

1. a) A molecule / compound [1]; that only contains carbon and hydrogen [1]
   b) The different molecules / hydrocarbons condense [1]; at a place in the column just below their boiling point [1]
2. O_2 [1]; 2 [1]

Page 48 Cracking Hydrocarbons

1. a) 2 [1]
   b) C_8H_{18} only [1]
2. Add bromine water [1]; if it goes colourless it is an alkene [1]; if it stays orange / brown / does not change colour then it is an alkane [1]

Page 49 Chemical Analysis

1. a) Paper [1]
   b) Pencil [1]; the ink from the pen would 'run' on the paper / dissolve in the solvent and affect the results [1]
   c) **B** and **C** [1]; there is only one spot [1]
   d) 18cm (distance moved by **X**) and 28cm (distance moved by solvent) [1]; $\frac{18}{28}$ [1] $= 0.64286 =$ 0.64 to 2 d.p. [1] (award 2 marks if calculation is correct but the answer is not given to 2 d.p.)

Page 50 The Earth's Atmosphere

1. C [1]
2. $\frac{0.05}{4.2} \times 100$ [1]; $= 1.2\%$ [1]
3. a) Evidence [1]; that supported their theory [1]
   b) **Any three of:** it happened a long time ago [1]; nobody was alive to record how it happened [1]; there is little evidence available [1]; there is evidence to support both theories [1]

Page 51 Greenhouse Gases

1. a) Carbon dioxide OR methane [1]; **plus:** increased burning of fossil fuels / increased deforestation (for carbon dioxide) OR more animal farming (digestion, waste decomposition) / decomposition of rubbish in landfill sites (for methane) [1]
   b) From: 0.6°C [1]; To: 3.6°C [1] (If no units are included only award 1 mark)
   c) **Any two of:** complex systems [1]; many different variables [1]; may only be based on parts of evidence [1]

Page 52 Earth's Resources

1. a) Water that is safe to drink [1]
   b) It contains dissolved substances [1]; it is a mixture as it contains more than just water molecules [1]
   c) **Any one of:** adding chlorine [1]; adding ozone [1]; using UV light [1]
   d) To remove the water from the salt [1]
   e) The sea water needs to be heated [1]; which requires a large amount of energy [1]

Page 53 Using Resources

1. **Any six of:** wood pulp is from trees, a renewable resource [1]; trees should be replanted before wood pulp can be considered a sustainable resource [1]; clay, chalk and titanium oxide are quarried which can have negative environmental impacts [1]; paper production uses a lot of water [1]; transportation of plastic bags uses less fuel [1]; plastic is longer lasting / can be reused many times [1]; (plastic longer lasting) so plastic may use less finite resources [1]; (plastic may use less finite resources) and so plastic may have a lower energy requirement [1]; paper is biodegradable so spends less time in landfill [1]; paper is more likely to be recycled which lowers raw material usage [1]

> For this type of question, you need to give your answer in a clear and logical way, using good English and correct grammar and punctuation. Typically, you might expect to receive 5–6 marks for a clear description of the advantages and disadvantages of both types of bag, with logical links; 3–4 marks if a number of relevant points are made, but the logic is unclear; 1–2 marks for fragmented points, with no logical structure.

Page 54 Forces – An Introduction

1. a) Weight / gravity [1]; non-contact [1]; air resistance (or drag) [1]; contact [1]
   b) weight = mass × gravitational field strength / $W = mg$ [1]; weight = $120\,000 \times 10 = 1\,200\,000$N [1]
   c) A correctly drawn horizontal arrow [1]; a vertical arrow that is half the length of the horizontal arrow [1]; and a diagonal arrow showing the total resultant force [1]

Resultant Force

2. a) Friction [1]
   b) Friction is a contact force [1]; lifting the glider off the track means there is no contact, so no friction [1]

Page 55 Forces in Action

1. a) Remove the weights [1]; if it returns to its original shape it was behaving elastically [1]
   b) i) 8.0 (cm) [1]
      ii) The extension appears to be linear [1]; and is increasing by 4cm each time [1]

c) force **[1]**; 0N **[1]**; 6N **[1]**
d) It allows you to check for errors / anomalies **[1]**; you can calculate a mean (average) result **[1]**

Page 56 Forces and Motion

1. a) Zero **[1]**

Displacement is the distance from the start point. The person returns home – it is a circular journey – so the total displacement at the end of their journey is zero.

b) B **[1]**
c) Stationary **[1]**
d) A is travelling away from home **[1]**; D is travelling in the opposite direction (back towards home) **[1]**; D is travelling slightly faster than A **[1]**

2. a) speed = $\frac{\text{distance travelled}}{\text{time}}$ / $v = \frac{s}{t}$ **[1]**; = $\frac{180}{6}$ = 30m/s **[1]**

b) i) 50 – 40 = 10m/s **[1]**
   ii) 50 + 40 = 90m/s **[1]**

The velocity changes from 50m/s to the east to 40m/s to the west.

Page 57 Forces and Acceleration

1. a) The mass of the system **[1]**
   b) The force accelerating the trolley (provided by the hanging masses) **[1]**
   c) The acceleration of the trolley **[1]**

The independent variable is the one deliberately changed and the dependent variable is the one being measured.

2. a) acceleration = $\frac{\text{change in velocity}}{\text{time taken}}$ / $a = \frac{\Delta v}{t}$ **[1]**; acceleration = $\frac{16 - 4}{8}$ = 1.5 **[1]**; m/s² **[1]**
   b) resultant force = mass × acceleration / $F = ma$ **[1]**; force = 68 × 1.5 = 102N **[1]**
3. a) acceleration = $\frac{33}{11}$ **[1]**; = 3m/s² **[1]**
   b) force = 950 × 3 **[1]**; = 2850N **[1]**

Page 58 Terminal Velocity and Momentum

1. They could change position e.g. dive head first **[1]**; to become more aerodynamic / reduce air resistance **[1]**
2. a) momentum = mass × velocity / $p = mv$ **[1]**; = 5000 × 1340 **[1]** = 6700000kg m/s **[1]**
   b) 6700000 in one second from 1 engine **[1]**; so 6700000 × 5 × 10 = 335000000kg m/s momentum to the exhaust gases **[1]**; and the

same amount of momentum is gained by the the rocket **[1]**

Momentum is conserved, so the momentum gained by the rocket is equal to the momentum given to the fuel.

c) momentum = mass × velocity / $p = mv$ **[1]**; 335000000 = 3000000 × v **[1]**; $v = \frac{335\,000\,000}{3\,000\,000}$ = 111.67m/s **[1]**

d) As the rocket burns fuel, it becomes lighter **[1]**; resultant force = mass × acceleration / $F = ma$, so acceleration (a) increases **[1]**; at a higher altitude, there is less air resistance **[1]**; so the resultant force increases **[1]**

Page 59 Stopping and Braking

1. a) 1.4 seconds **[1]**
   b) 1.4 × 15 **[1]**; = 21m **[1]**
   c) 4 – 1.4 = 2.6 seconds **[1]**
   d) momentum = mass × velocity / $p = mv$ **[1]**; 1300 × 15 = 19500kg m/s **[1]**
   e) It would be longer **[1]**
   f) A correctly drawn graph line that starts horizontally at 15m/s **[1]**; then starts sloping downwards between 0.2s and 0.8s **[1]**; and has the same gradient on the downslope as the original line **[1]**

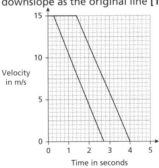

g) **Any two of:** The down slope would start at the same point **[1]**; but have a shallower gradient **[1]**; and take a longer total time to stop **[1]**

Page 60 Energy Stores and Transfers

1. a) 100 – 20 = 80 degree change **[1]**; energy = 2 × 4200 × 80 **[1]** = 672000J **[1]**
   b) 672000 = 2.5 × 4200 × temp change **[1]**; temp change = $\frac{672\,000}{(2.5 \times 4200)}$ = 64°C **[1]**; final temp = 20 + 64 = 84°C **[1]**;
2. a) gravitational potential energy = mass × gravitational field strength × height / $E_p = mgh$ **[1]**; E_p = 0.1 × 10 × 0.05 **[1]**; = 0.05J **[1]**

b) kinetic energy = 0.5 × mass × (speed)² / $E_k = \frac{1}{2}mv^2$ **[1]**; 0.05 = 0.5 × 0.1 × v^2 **[1]**; $v^2 = \frac{0.05}{(0.5 \times 0.1)}$ = 1, v = 1m/s **[1]**

Page 61 Energy Transfers and Resources

1. light **[1]**; electrical / thermal **[1]**; heat **[1]**
2. a) Start temperature of water **[1]**; thickness of fleece **[1]**
   b) Fleece M **[1]**; because it cools the slowest, so has the lowest thermal conductivity **[1]**; and insulates the best **[1]**

Page 62 Waves and Wave Properties

1. a) Half a wave per second **[1]** (Accept: 1 wave every 2 seconds)
   b) wave speed = frequency × wavelength / $v = f\lambda$ **[1]**
   c) speed = $\frac{\text{distance}}{\text{time}}$ / $v = \frac{s}{t}$, $v = \frac{50}{10}$ **[1]**; = 5m/s **[1]**
   d) $v = f\lambda$, $\lambda = \frac{5}{0.5}$ **[1]**; = 10m **[1]**
   e) They will go slower **[1]**
2. In longitudinal waves, the particles oscillate **[1]**; parallel to the direction of energy transfer / wave motion **[1]**; in transverse waves, the oscillation is at right-angles to the direction of energy transfer / wave motion **[1]**

Page 63 Electromagnetic Waves

1. Can be shown on the diagram to help explain but must include 'refraction' in the answer for full marks, e.g. Light rays from the pin **[1]**; are refracted when they leave the water **[1]**; away from the normal and into the eye **[1]**

2. frequency **[1]**; wavelength **[1]**

Page 64 The Electromagnetic Spectrum

1. a) Microwaves **[1]**
   b) Accept any sensible answer, e.g. X-rays for photographing bones OR gamma rays for sterilisation OR UV for sunbeds **[2]** (1 mark for wave, 1 mark for use)
2. a) X-rays **[1]**
   b) It can penetrate soft tissue **[1]**; but is blocked by bone **[1]**
   c) i) Skin cancer **[1]**
      ii) It has more / higher frequency energy **[1]**; and is ionising **[1]**

Answers

iii) **Any two of:** they think it looks good / healthy [1]; they don't think it will happen to them [1]; they don't think it is that risky [1] (Accept any other sensible answer)

Page 65 An Introduction to Electricity

1. current [1]; charge [1]; greater [1]; current [1]
2. a) potential difference = current × resistance / $V = IR$ [1]
   b) $230 = 5 \times R$ [1]; $R = \frac{230}{5} = 46\Omega$ [1]
   c) charge = current × time / $Q = It$ [1]
   d) $Q = 5 \times 120$ [1]; = 600C [1]
3. energy [1]; greater [1]; current [1]
4. a) Closed switch [1]
   b) Battery [1]
   c) Fuse [1]
   d) Light dependent resistor / LDR [1]

Page 66 Circuits and Resistance

1. a) To adjust the resistance of the circuit [1]; and change the voltage across the component [1]
   b) Series [1]
   c) Parallel [1]
2. Four correctly drawn lines [3] (2 marks for two correct lines; 1 mark for one correct line)
   Light dependent resistor (LDR) – Resistance decreases as light intensity increases.
   Thermistor – Resistance decreases as temperature increases.
   Diode – Has a very high resistance in one direction.
   Filament light – Resistance increases as temperature increases.

Page 67 Circuits and Power

1. a) energy transferred = power × time / $E = Pt$ [1]; $E = 1600 \times 120$ [1]; = 192 000J [1]
   b) $\frac{192\,000}{100} \times 10$ [1]; = 19 200J [1]
   (accept any equivalent method)
2. a) $V = IR$, $V = 2 \times 3$ [1]; = 6V [1]
   b) $18 = 6 + V$ [1]; $V = 18 - 6 = 12$V [1]

> The total potential difference in a series circuit is shared across the components.

   c) $18 = 2 \times R$ [1]; $R = \frac{18}{2} = 9\Omega$ [1]

Page 68 Domestic Uses of Electricity

1. a) 1.5V [1]; d.c. [1]
   b) 230V [1]; a.c. [1]
2. When a device is switched off, the live wire before the switch can still be at a non-zero potential [1]; touching this would create a potential difference between the wire and the ground [1]; this would make current flow through the person [1]; which would cause an electric shock [1]

Page 69 Electrical Energy in Devices

1. a) Kinetic [1]
   b) It is dissipated / lost [1]; to the surroundings [1]
2. a) Energy input = electrical [1]; useful energy output = kinetic [1]
   b) It disappears. [1]
   c) 500J [1]
3. The output from the generators goes through a step-up transformer [1]; this increases the voltage and also reduces the current [1]; the low current stops the cables from becoming hot [1]; which means less energy is lost during transmission [1]

Page 70 Magnetism and Electromagnetism

1. If free to move, the magnet will rotate so that the north pole of the magnet [1]; points to the Earth's north pole [1]
2. a) They will repel [1]
   b) It will be attracted to the magnet [1]
3. When the switch is pressed current flows in the electromagnet [1]; this magnetises the magnet [1]; which attracts the armature, causing the hammer to hit the gong [1]; the movement of the armature breaks the circuit, switching off the magnet [1]; the armature springs back and remakes the circuit, which starts the cycle again [1]

Page 71 The Motor Effect

1. a) force on a conductor (at right-angles to a magnetic field carrying a current = magnetic flux density × current × length / $F = BIl$,
   $F = (0.3 \times 10^{-3}) \times 2 \times 0.05$ [1];
   = 0.00003N [1]
   b) There will be no force on the wire [1]
2. a) 1st finger = direction of field [1]; 2nd finger = direction of current [1]; thumb = direction of force [1]
   b) Increase the current [1]; use stronger magnets [1]
   c) Reverse the current [1]; reverse the magnets / magnetic field [1]

Page 72 Particle Model of Matter

1. a) The energy required [1]; to change 1kg of a substance from a solid to a liquid [1]
   b) thermal energy for a change of state = mass × specific latent heat / $E = mL$, $E = 0.012 \times (2.3 \times 10^{6})$ [1]; = 27 600J [1]
2. a) The substance is condensing [1]; from a gas to a liquid [1]
   b) The particles are slowing down [1]; and the substance is cooling [1]

Page 73 Atoms and Isotopes

1. a) Electron, –1 [1]; neutron, 0 [1]; proton, +1 [1]
   b) It has the same number of protons [1]; as electrons [1]
   c) ion [1]; positive [1]
2. Path **A** is a long way from the nucleus and the alpha particle goes straight through [1]; Path **B** is close to the positive nucleus so the alpha particle is deflected [1]; Path **C** comes very close to the nucleus and the alpha particle is repelled back the way it came [1]

> Two positively charged particles will repel each other.

Page 74 Nuclear Radiation

1. a) An unstable atom that gives out radiation [1]
   b) Beta decay [1]
   c) It is stable / non-radioactive [1]
   d) Sodium [1]
2. Becquerel [1]
3. Gamma, beta, alpha [1]

Page 75 Half-Life

1. a) No longer a risk [1]; because it has a half-life of just 8 days [1]; so would have completely decayed to the same level as background radiation in the last 30 years [1]
   b) 24 days is 3 half-lives [1]; $256 \rightarrow 128 \rightarrow 64 \rightarrow 32$, so a count rate of 32 remains [1]
   c) To fall to $\frac{1}{8}$ takes 3 half-lives, so $(3 \times 30 =)$ 90 years [1]; $(1986 + 90 =)$ 2076 [1]
   d) $^{40}_{19}$potassium [1]; $\rightarrow ^{40}_{20}$caesium $+ ^{0}_{-1}$e [1]

Pages 77-92 Biology Practice Exam Paper 1

01.1 Bacterium [1]
01.2 $\frac{70}{14\,000}$ [1]; = 0.005 [1]
01.3 2, 4, 8, 16, 32, 64 (idea of doubling) [1]; 64 [1]
02.1 **Any two of:** fever [1]; vomiting [1]; abdominal cramps [1]; diarrhoea [1]
02.2 Incidences higher in the summer [1]; because food not kept at a cold enough temperature in summer / references to barbeques and undercooked food [1]
03.1 Right ventricle [1]
03.2 A build-up of fatty material inside the coronary arteries [1]; narrows them down and reduces the flow of blood [1]; resulting in a lack of oxygen for the heart muscle [1]; **Any three risk factors from:** saturated fat in diet [1]; smoking [1]; lack of exercise [1]; stress [1]; genetic factors [1]
03.3 Will have the same antigens / tissue type [1]; so no chance of rejection [1]

03.4 Stem cells removed from embryos [1]; the embryos are destroyed / mention of ethics of disposing of human embryos [1]

04.1 Fleming [1]; fungus [1]

04.2 4, 3, 1, 2 [3] (1 mark for 4 before 3; 1 mark for 3 before 1; 1 mark for 1 before 2)

05.1 To outcompete other plants [1]; and get more light [1]; so more photosynthesis / they can produce more food [1]

05.2 Water is transported in the xylem [1]; it is pulled up [1]; due to loss of water / transpiration from the leaves [1]

05.3 The presence of spines [1]; mean animals are less likely to eat it [1]; plus lower numbers of stomata [1]; so less water loss [1]; the thick waxy cuticle [1]; means less loss of water through the epidermis [1]

06.1 Correctly plotted points [3]; straight line of best fit [1]

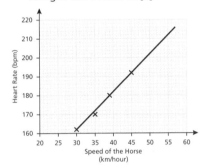

06.2 As speed increases heart rate increases [1]; Straight line (for these speeds) / use of figures [1]

06.3 A horizontal line should be drawn on the graph from 200bpm on the *y*-axis to the line of best fit and a vertical line should be drawn down from this point to the *x*-axis [1]; accept 48, 49 or 50 km/hour [1]

06.4 glucose [1]; → lactic acid [1]

06.5 No lactic acid is made [1]; so muscles are not fatigued [1]; and more energy released [1]

07.1 Fatty acids are formed [1]

07.2 Bile is present [1]; which emulsifies the fats [1]; giving a larger surface area for lipase to work on [1]

07.3 The lipase had been boiled [1]; so the enzyme had been denatured [1]

07.4 Amylase only works on starch or carbohydrates / protease only works on protein [1]; lipase is the wrong shape to work on starch / protein [1]; reference to the lock and key model and specific fit [1]

08.1 The movement of water [1]; from a dilute solution to a concentrated solution [1]; through a selectively permeable membrane [1]

08.2 The cylinders had lost water [1]; by osmosis [1]; because the contents of the potato cells were more dilute than the solution [1]

08.3 Answer from intercept on graph in the range of 0.35–0.38mol/dm³ [1]

08.4 **Top to bottom: I [1]; C [1]; D [1]; C [1]**

Pages 93-108 Biology Practice Exam Paper 2

01.1 Fingertip [1]

01.2 To make sure the results are reliable [1]; because she has a 50/50 chance of guessing right [1]

01.3 Less sensitive than fingertip / more sensitive than leg [1]; she can tell the difference between one pin and two at 1cm only [1]

01.4 She is thinking about it / it is not involuntary [1]; it does not involve protecting her body [1]

01.5 **Top to bottom:** (A), D, E, B, C [3] (2 marks for two correct; 1 mark for one correct)

02.1 **Any two of:** sharp teeth to catch seals [1]; white fur for camouflage [1]; sharp claws to catch seals [1]; forward facing eyes to judge distance [1]

02.2 They are in the same genus (*Ursus*) [1]; but are in different species [1]

02.3 Polar bears and Alaskan bears may mate and produce sterile hybrids. [1]; The habitats of the polar bears and the Alaskan bears may overlap. [1]

02.4 Increased burning of fossil fuels [1]; and deforestation are causing the increase [1]; carbon dioxide is a greenhouse gas [1]; it traps long wavelength energy from the sun [1]; that is being reflected back by the Earth's surface [1]; leading to an increase in temperature [1]

03.1 Copper from waste is passing into the river [1]; the concentration is low before the waste pile / high directly after the waste pile / decreases in concentration as you move away from the waste pile [1]

03.2 At **B** the concentration of copper is higher [1]; plants at **B** are being poisoned by the copper [1]

03.3 Plants at **B** have adapted to living in high copper concentrations [1]; plants at **A** have not experienced high concentrations before [1]

03.4 All plants have a different resistance to copper [1]; the ones that are most resistant survive [1]; they reproduce and pass on the genes for resistance [1]; and over many generations the plants become more resistant [1]

04.1 heterozygous [1]; tall [1]; dominant [1]; genotypes [1]

04.2 1:1 [1]; (Accept a final answer of 50% or $\frac{1}{2}$)

	T	t	
	T	t	[1];
t	Tt	tt	[1];
t	Tt	tt	[1]

04.3 **Any two of:** two Tt plants could produce 10 tall plants [1]; it is unlikely, but could happen by chance [1]; 10 plants is not enough, more crosses need to be done [1]

05.1 If Y chromosome is present then it is a boy [1]; as boys have XY sex chromosomes [1]

05.2 Easier to obtain mother's blood than baby's cells [1]; less risk / less likely to cause damage or miscarriage [1]

06.1 Asexual [1]

06.2 They have exactly the same genes [1]

06.3 The bottom part contained the nucleus [1]; this means in has the genetic material / DNA / chromosomes [1]; and can make new proteins [1]

07.1 15 + 21 = 36% [1]

07.2 Irregular ovulation [1]; 75% of 16 is the highest figure [1]

07.3 If levels of progesterone increase less FSH is released / if levels of progesterone fall then more FSH is released [1]

07.4 Less progesterone is released [1]; so more FSH is released [1]; and eggs are stimulated to develop [1]

08.1 A chemical messenger [1]; carried in the blood [1]

08.2 **From top to bottom:** ADH [1]; pituitary gland [1]; thyroid [1]; controls metabolic rate [1]; adrenaline [1]; gets the body ready for action [1]

08.3 **Any four of:** they cannot produce insulin [1]; therefore they cannot reduce their glucose levels [1]; glucose may start to pass out in the urine [1]; high levels can lead to a coma [1]; they need to know to inject themselves with insulin [1]

Pages 109–124 Chemistry Practice Exam Paper 1

01.1 Proton +1, Neutron 0, Electron –1 [1]

01.2 It has equal numbers of protons and neutrons / equal numbers of positive and negative charges [1]

01.3 Mass number 4, Atomic number 2 [1]

Answers

01.4 It has a full outer shell of electrons. **[1]**

01.5 It has the same atomic number. **[1]**; It has a different mass number. **[1]**

02.1 5 **[1]**

02.2 (12 + 16 + 16) = 44 **[1]**

02.3 oxidation–carbon **[1]**; reduction–iron(III) oxide **[1]**

02.4 It is found pure in the ground **[1]**; because it is unreactive **[1]**

03.1 **Any one of:** the density increases as you go down the group **[1]**; the melting point increases as you go down the group **[1]**; the boiling point increases as you go down the group **[1]** (Accept alternative answers, e.g. the density decreases as you go up the group)

03.2 Any answer in the range 200–230°C **[1]**

03.3 One shared pair of electrons drawn between the two atoms **[1]**; three non-bonding pairs drawn on each atom **[1]** (the second mark will only be awarded if the first mark is achieved)

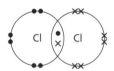

03.4 Br_2 **[1]**

03.5 Intermolecular forces break **[1]**

03.6 They all have the same number of electrons (seven) in their outer shell **[1]**

03.7 Sodium ion has no electrons drawn **[1]**; labelled Na^+ / + ion **[1]**; chlorine ion has eight electrons drawn **[1]**; seven represented by dots and one a cross **[1]**; labelled Cl^- / 1– ion **[1]**

03.8 The atoms get larger **[1]**; the outer shell gets further from the nucleus **[1]**; the attraction between the nucleus and electrons gets weaker **[1]**; so an electron is less easily gained **[1]**

04.1 Filtration **[1]**; to remove the excess copper(II) oxide **[1]**

04.2 Hazard: **any one of:** chemical on skin **[1]**; chemical in eyes **[1]**; cuts from broken glass **[1]**;
Way of reducing the risk: **any one of (as appropriate to hazard):** wash hands **[1]**; wear eye protection **[1]**; ask teacher to clear away broken glass **[1]**

04.3 M_r CuO = 63.5 + 16 = 79.5 **[1]**;
M_r $CuSO_4.5H_2O$ = 63.5 + 32 + (16 × 4) + 5 × (2 × 1 + 16) = 249.5 **[1]**;
moles CuO = $\frac{5.2}{79.5}$ = 0.0654088 **[1]**;
mass $CuSO_4.5H_2O$ =
0.0654088 × 249.5 = 16.3g **[1]**

05.1 Similarity: **any one of:** they both are made up of carbon atoms **[1]**; both contain strong covalent bonds between carbon atoms **[1]**
Difference: **any one of:** graphite is made up of layers, diamond has no layers **[1]**; graphite contains weak intermolecular forces, diamond only contains strong covalent bonds **[1]**; graphite has delocalised (free) electrons, diamond does not **[1]**

05.2 Electrical conductivity comes from delocalised electrons, which are able to move through the structure **[1]**; this is useful for touch-screens, as they need to be able to conduct electricity to work **[1]**; strength comes from strong covalent bonds between carbon atoms **[1]**; this is useful for touch-screens so they do not crack / shatter when dropped **[1]**; graphene is transparent because it is only one atom thick **[1]**; this is useful for touch-screens so you can see the light coming through from the display underneath **[1]**

06.1 Battery / cell added and joined to electrodes **[1]**; electrode connected to positive side of cell labelled anode **[1]**; electrode connected to negative side of cell labelled cathode **[1]**

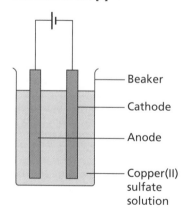

— Beaker

— Cathode

— Anode

— Copper(II) sulfate solution

06.2 Copper **[1]**; it is less reactive than hydrogen **[1]**

06.3 2 **[1]**; 4 **[1]**

06.4 Oxidation **[1]**; because electrons are lost from the (hydroxide) ions **[1]**

07.1 **In order:** (s), (aq), (g) **[1]**

07.2 To let the gas out / to stop the acid spraying **[1]**

07.3 Sensible scales, using at least half the grid for the points **[1]**; all points correct **[2]**; connected by smooth curve **[1]**

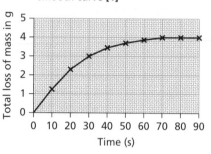

07.4 70s **[1]**

07.5 All of the acid had reacted **[1]**

07.6 Hydrogen gas is formed **[1]**; which escapes into the air **[1]**

07.7 Less steep line to right of original line **[1]**; finishes at same overall mass loss **[1]**

08.1 Bonds broken = (432 × 2) + 495 = 1359kJ/mol **[1]**; bonds formed = 467 × 4 = 1868kJ/mol **[1]**; bonds broken – bonds formed = –509kJ/mol **[1]**

08.2 Less energy is required to break the bonds in the reactants **[1]**; than is produced when the bonds are formed in the products **[1]**

Pages 125–140 Chemistry Practice Exam Paper 2

01.1 2 **[1]**

01.2 carbon particles **[1]**

01.3 It is colourless **[1]**; and odourless **[1]**

01.4 $N_2 + 2O_2 \rightarrow 2NO_2$ **[1]**

01.5 sulfur dioxide – acid rain **[1]**; carbon particles – global dimming **[1]**

02.1 Formula: C_3H_6 **[1]**; Name: Propene **[1]**

02.2 Add bromine water **[1]**; turns orange to colourless **[1]**

02.3 HD poly(ethene) **[1]**; because it is stronger **[1]**

03 **Effects on the environment: any two of:** melting of ice caps **[1]**; sea level rise, which may cause flooding and coastal erosion **[1]**; changes in

amount, timing and distribution of rainfall **[1]**; desertification in some regions **[1]**; more frequent and severe storms **[1]**

Effects on people: any two of: flooding of homes **[1]**; migration of people from affected areas **[1]**; temperature and water stress **[1]**; lack of food in some regions **[1]**

Effects on wildlife: any two of: changes to distribution of species **[1]**; extinction of some species **[1]**; temperature and water stress **[1]**; lack of food in some regions **[1]**

04.1 Desalination / distillation **[1]**

04.2 It contains other substances **[1]**; which are dissolved **[1]** (Accept 'it's a mixture' for 1 mark)

04.3 Heat the water (with the Bunsen burner) **[1]**; if the water is pure it will start to boil at 100°C **[1]**

04.4 Chlorine / ozone / ultraviolet light **[1]**

04.5 $\frac{250}{1000} \times 1.35$ **[1]**; = 0.3375 **[1]**; = 0.34mg (to 2 decimal places) **[1]**

05.1 They are finite / non-renewable / may run out **[1]**

05.2 **Any five valid points, e.g.** recycling reduces the amount of metal being mined (metals will last longer) **[1]**; mining, processing metals and recycling all require energy, which mostly comes from the use of finite resources **[1]**; collecting and transporting cans uses petrol / diesel **[1]**; sorting cans and rolling metal blocks requires electrical machinery **[1]**; melting the cans requires a lot of energy **[1]**; overall, recycling consumes less energy than producing new cans **[1]**

06.1 Push the plunger of the syringe down to remove any air in it **[1]**; the plunger may be pushed out before the end of the reaction **[1]**; so volume of gas produced not accurately measured **[1]**

06.2 All points plotted correctly for two sets of data **[2]**; two lines of best fit drawn **[2]**

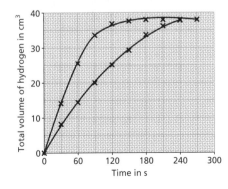

06.3 The higher the concentration, the faster the rate of reaction **[1]**

06.4 As the concentration increases so does the number of particles of acid in a given volume **[1]**; so there are more frequent collisions / more collisions per second with magnesium particles **[1]**; so rate increases / reaction speeds up **[1]**

06.5 140s **[1]**

06.6 Tangent drawn at 50s **[1]**; volume of hydrogen calculated, e.g. 0.64 – 0.36 = 0.28 **[1]**; time calculated, e.g. 60 – 30 = 30 **[1]**; gradient calculated to give rate of reaction, e.g. $\frac{0.28}{30}$ = 0.009333 **[1]**; = 0.009 **[1]**; cm³/s **[1]** (Accept 0.008)

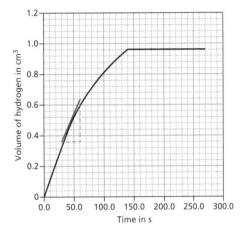

07.1 2.5 billion years **[1]**

07.2 40cm³ **[1]**

07.3 Algae evolved and starting producing oxygen **[1]**; via photosynthesis **[1]**; land plants evolved to increase amounts of oxygen in the atmosphere further **[1]**

07.4 Carbon dioxide allows short wavelength radiation to pass through **[1]**; the atmosphere to the Earth's surface **[1]**; carbon dioxide absorbs outgoing long wavelength radiation **[1]**

07.5 Increased **[1]**

07.6 Increased burning of fossil fuels **[1]**; in vehicle engines/power stations **[1]** OR increased deforestation **[1]**; so fewer trees to absorb carbon dioxide from the air **[1]**

08.1 3 **[1]**

08.2 15cm (distance moved by S) and 22cm (distance moved by solvent) **[1]**; $\frac{15}{22}$ = 0.68182 **[1]**; = 0.68 **[1]**

08.3 Mobile phase / solvent moves through the paper **[1]**; and carries different compounds different distances **[1]**; depending on their attraction for the paper and the solvent **[1]**

Pages 141–154 Physics Practice Exam Paper 1

01.1 From left to right: 65J **[1]**; 5J **[1]**; 30J **[1]**

01.2 Heat the schools **[1]**; because it saves the most energy **[1]**; half of 5J is 2.5J **[1]**; but a quarter of 65J is over 15J **[1]**

01.3 efficiency = $\frac{\text{useful output energy transfer}}{\text{total input energy transfer}}$ ×100%, = $\frac{30}{100} \times 100\%$ = 30% **[1]**

02.1 It spreads out / is dissipated **[1]**; into the surroundings **[1]**

02.2 20 + 13 + 7 = 40% **[1]**

02.3 60% **[1]**

03.1 work done = force × distance / $W = Fs$, $W = 6 \times 2$ **[1]**; = 12J **[1]**

03.2 weight = mass × gravitational field strength / $W = mg$, $m = \frac{6}{10} = 0.6$kg **[1]**; gravitational potential energy = mass × gravitational field strength × height / $E_p = mgh$, $E_p = 0.6 \times 10 \times 2 = 12$J **[1]**

03.3 kinetic energy = 0.5 × mass × (speed)² / $E_k = \frac{1}{2}mv^2$ **[1]**; $v^2 = \frac{12}{0.5 \times 0.6} = 40$ **[1]**; $v = \sqrt{40}$ = 6.3m/s **[1]**

04.1 Top to bottom: gravitational **[1]**; light **[1]**; kinetic **[1]**; chemical **[1]**

04.2 Coal and gas **[1]**

04.3 Nuclear / oil **[1]**

04.4 Wave / tidal / geothermal / biomass **[1]**

04.5 Nuclear power stations have high output **[1]**; and are reliable **[1]**; **plus any two of:** however they produce nuclear waste **[1]**; are expensive to build and decommission **[1]**; have a risk of explosion if something went wrong **[1]**; and are not renewable **[1]**

05.1 **Y** = variable resistor **[1]**; **Z** = voltmeter **[1]**

05.2 To control the voltage applied to component **X** / adjust the resistance **[1]**

05.3 The voltage **[1]**

05.4 The current **[1]**

05.5 Accurately plotted voltage **[1]**; and current **[1]**; with all points connected **[1]**

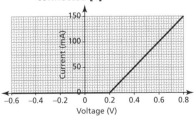

Answers

05.6 Yes, they are correct as all the points fit the line **[1]**

An anomalous result would be significantly higher or lower than the other results or would not fit the pattern.

05.7 A diode **[1]**
06.1 They have the same number of protons **[1]**
06.2 Beta decay **[1]**
06.3 The mass number has not changed during emission, but the proton number has increased by one **[1]**
06.4 The time it takes for half of the radioactive isotopes to decay / for the count rate to halve **[1]**
06.5 $\frac{1}{8}$ means that 3 half-lives have passed **[1]**; so is (30 × 3 =) 90 years **[1]**
07.1 kinetic energy = 0.5 × mass × (speed)^2 /
$E_k = \frac{1}{2}mv^2$, $m = \frac{E_k}{\frac{1}{2}v^2}$ **[1]**;
$m = \frac{472500}{450}$ **[1]**; = 1050 **[1]**; final answer = 1050kg (unit must be correctly stated for mark) **[1]**
07.2 The bus has more mass **[1]**; so it has more kinetic energy **[1]**
07.3 kinetic energy = 0.5 × mass × (speed)^2 / $E_k = \frac{1}{2}mv^2$ **[1]**; increase in E_k = 0.5 × 1200 × (14^2 – 8^2) **[1]**; = 79 200J **[1]**
07.4 Because the kinetic energy depends on the square of the speed **[1]**
08.1 gravitational potential energy = mass × gravitational field strength × height / $E_p = mgh$,
E_p = 65 × 10 × 1.25 **[1]**; = 812.5J **[1]**
08.2 kinetic energy =
0.5 × mass × (speed)^2 / $E_k = \frac{1}{2}mv^2$,
812.5 = 0.5 × 65 × v^2 **[1]**; v^2 = 25 **[1]**;
$v = \sqrt{25}$ = 5m/s **[1]**
09.1 12V **[1]**
09.2 2 + 2 = 4A **[1]**
09.3 potential difference = current × resistance / $V = IR$,
$R = \frac{12}{2}$ **[1]**; = 6Ω **[1]**
09.4 It is less / half **[1]**
09.5 power = potential difference × current / $P = VI$, P = 12 × 2 **[1]**; = 24W **[1]**
09.6 $\frac{96\,000}{(24 \times 2)}$ **[1]**; = 2000s **[1]**

Pages 155–168 Physics Practice Exam Paper 2

01.1 So it will be attracted by the magnetic coil **[1]**
01.2 It will need to increase **[1]**
01.3 A bigger mass makes a bigger force pulling down on the left, so a bigger force is needed on the right **[1]**; a bigger current will increase the strength of the magnetic field created by the coil **[1]**
01.4 It will increase the mass that can be balanced **[1]**; because the iron core will make the magnet stronger **[1]**
02.1 Towards the centre of the circle **[1]**
02.2 The tension of the string **[1]**
02.3 The radius of the circle **[1]**; and the mass of the bung **[1]**
02.4 speed **[1]**; velocity **[1]**; direction **[1]**
02.5 Gravity / the gravitational attraction of the Earth on the Moon **[1]**
03.1 **From top to bottom:** 1500m, 3 × 10^{-8}m, 5 × 10^{-7}m, 1 × 10^{-11}m **[3]** (2 marks for two in the correct position; 1 mark for one correct)
03.2 **Any two of:** wave speed **[1]**; can travel through a vacuum **[1]**; transverse waves **[1]**
03.3 **Any pair from:** infrared **[1]**; used for cooking / remote control **[1] OR** Microwaves **[1]**; used for communication / cooking **[1] OR** gamma rays **[1]**; used for sterilising equipment **[1]**
03.4 wave speed = frequency × wavelength / $v = f\lambda$ **[1]**;
$f = \frac{3 \times 10^8}{300}$ **[1]**; = 1 × 10^{6}Hz **[1]**
03.5 acceleration = $\frac{\text{change in velocity}}{\text{time}}$ /
$a = \frac{\Delta v}{t}$ **[1]**; $a = \frac{10}{4}$ **[1]**; = 2.5m/s^2 **[1]**
03.6 10m/s **[1]**
03.7 distance travelled = speed × time / $s = vt$ **[1]**; s = 10 × 4 = 40m **[1]**
04.1 speed = $\frac{\text{distance travelled}}{\text{time}}$ / $v = \frac{d}{t}$
[1]; = $\frac{(350 - 100)}{(7 - 3)}$ **[1]**; = 6.5m/s **[1]**;
the car travels at constant speed for the first 3 seconds **[1]**; it then speeds up to 6.5m/s for 4 seconds **[1]**; before slowing again for the next 3 seconds **[1]**

To gain full marks in this type of question, your response must be written using correct grammar and punctuation. Your ideas must in a sensible order and use the appropriate scientific words.

04.2 acceleration = $\frac{\text{change in velocity}}{\text{time taken}}$ /
$a = \frac{\Delta v}{t}$ **[1]**; = $\frac{30}{6}$ **[1]**; = 5m/s^2 **[1]**
05.1 resultant force = mass × acceleration / $F = ma$ **[1]**;
F = 3500 – 2000 = 1500 **[1]**;
$a = \frac{1500}{1200}$ **[1]**; = 1.25 m/s^2 **[1]**
05.2 As the speed increases, the resistive forces increase **[1]**; until they balance the engine thrust / it reaches terminal velocity **[1]**
05.3 The distance a car travels between the driver seeing the hazard and applying the brakes **[1]**
05.4 **Any two of:** fatigue **[1]**; age **[1]**; alcohol / drugs **[1]**; distractions **[1]** (Accept named distractions, e.g. using a mobile phone)
05.5 Road conditions (Accept a specific adverse condition, e.g. ice, snow, rain, etc.) **[1]**
05.6 As the brakes are applied friction occurs between the brakes and the wheels **[1]**; this takes kinetic energy from the car, which is converted to heat **[1]**
06.1 Wavelength **[1]**
06.2 Amplitude **[1]**
06.3 Sound **[1]**
06.4 wave speed = frequency × wavelength / $v = f\lambda$ **[1]**;
v = 8 × 0.015 **[1]**; = 0.12m/s **[1]**

8 waves per second means 8Hz.

06.5 Set up the tank and switch on the stroboscope and adjust the scopes flashing speed so that the waves (or shadows of the waves) appear stationary **[1]**; measure the wavelength with a ruler **[1]**; if using shadows, scale the value based on the total size of shadow compared to the total size of tank **[1]**; flashing lights are a hazard as they can trigger photosensitive epilepsy **[1]**
07.1 They are reversed **[1]**
07.2 They would be closer together / more of them in the same space **[1]**
07.3 First finger: field **[1]**; second finger: current **[1]**; thumb: movement / force **[1]**
07.4 Into the paper / downwards at 90° to the surface of the page **[1]**
07.5 **Any three of:** a weaker current **[1]**; using weaker magnets **[1]**; moving the magnets further apart **[1]**; changing the angle of the wire so not at right-angles to the field **[1]**

Notes

Acknowledgements

The author and publisher are grateful to the copyright holders for permission to use quoted materials and images.

Cover & p1 watchara/Shutterstock.com; cover & p1 everything possible/Shutterstock.com; cover & p1 Ahuli Labutin/Shutterstock.com; p7 D. Kucharski K. Kucharska/Shutterstock.com; p8 STEVE GSCHMEISSNER/Science Photo Library; p84 Perfect Lazybones/Shutterstock.com; p85 scenery2/Shutterstock.com; p86 Nadezda Murmakova/Shutterstock.com; p86 Pavel L Photo and Video/Shutterstock.com; p86 Marques/Shutterstock.com; p96 Jim David/Shutterstock.com; p100 National Library Of Medicine/Science Photo Library

Every effort has been made to trace copyright holders and obtain their permission for the use of copyright material. The author and publisher will gladly receive information enabling them to rectify any error or omission in subsequent editions. All facts are correct at time of going to press.

Published by Collins
An imprint of HarperCollinsPublishers Ltd
1 London Bridge Street
London SE1 9GF

© HarperCollinsPublishers Limited 2016

ISBN 9780008326678

Content first published 2016
This edition published 2018

10 9 8 7 6 5 4 3 2 1

British Library Cataloguing in Publication Data.

A CIP record of this book is available from the British Library.

Commissioning Editor: Emily Linnett and Fiona Burns
Biology author: Ian Honeysett
Chemistry authors: Emma Poole and Gemma Young
Physics author: Nathan Goodman
Project Manager: Rebecca Skinner
Project Editor: Hannah Dove
Designers: Sarah Duxbury and Paul Oates
Copy-editor: Rebecca Skinner
Proofreader: Aidan Gill
Typesetting and artwork: Jouve India Private Limited
Production: Lyndsey Rogers
Printed by Martins the Printers